产教融合信息技术类“十三五”规划教材
西普教育研究院 IT 前沿技术方向校企合作系列教材

“十三五”江苏省高等学校重点教材
（编号：2019-2-037）

U0191229

数据采集与预处理

Data Acquisition and Preprocessing

米洪 张鸰 ◎ 主编
季丹 史律 张俊 ◎ 副主编

人 民 邮 电 出 版 社
北 京

图书在版编目（CIP）数据

数据采集与预处理 / 米洪，张鸽主编. -- 北京 : 人民邮电出版社，2019.11（2024.6重印）
产教融合信息技术类“十三五”规划教材
ISBN 978-7-115-51915-3

Ⅰ. ①数… Ⅱ. ①米… ②张… Ⅲ. ①数据采集－高等学校－教材②数据处理－高等学校－教材 Ⅳ. ①TP274

中国版本图书馆CIP数据核字(2019)第190750号

内容提要

本书以任务驱动为主线，围绕企业级应用进行项目任务设计，主要内容包括数据采集与预处理准备、网络爬虫实践、日志数据采集实践和数据预处理实践，全面地讲述了 Scrapy、Flume、Pig、Kettle、Pandas、OpenRefine 等技术，以及 urllib、Selenium 基本库和 BeautifulSoup 解析库的相关知识与应用案例。

本书内容实用，可操作性强，语言精练，通俗易懂，可作为高等院校计算机应用技术、大数据技术与应用、软件技术、云计算技术与应用等计算机相关专业的教材，也可作为大数据分析、云计算应用领域技术人员的参考用书。

◆ 主　　编　米　洪　张　鸽
副 主 编　季　丹　史　律　张　俊
责任编辑　左仲海
责任印制　王　郁　马振武
◆ 人民邮电出版社出版发行　　北京市丰台区成寿寺路 11 号
邮编　100164　　电子邮件　315@ptpress.com.cn
网址　http://www.ptpress.com.cn
山东百润本色印刷有限公司印刷
◆ 开本：787×1092　1/16
印张：11.5　　2019 年 11 月第 1 版
字数：381 千字　　2024 年 6 月山东第 10 次印刷

定价：39.80 元

读者服务热线：(010)81055256　印装质量热线：(010)81055316
反盗版热线：(010)81055315
广告经营许可证：京东市监广登字 20170147 号

前言 FOREWORD

1. 缘起

大数据技术在人们的日常生活中已经被广泛应用，信息技术与经济社会的交汇融合引发了数据的迅猛增长，数据已成为国家基础性战略资源。大数据正日益对全球生产、流通、分配、消费、经济运行机制和社会生活方式产生重大影响。大数据技术通过对数量巨大、来源分散、格式多样的数据进行采集、存储和关联分析，从中发现新知识、创造新价值、提升新能力，已经发展成为新一代的信息技术和服务产业。

大数据应用的关键是“数据”，“机器学习”“通用算法”的崛起使得“数据”的地位更加凸显。然而，各行各业的信息化建设都是封闭式的，海量数据被封装在不同的软件系统中，数据源多种多样，且更新较快。如何采集这些数据资源，如何对采集的数据进行预处理，去除其中的“脏数据”，是大数据分析面临的问题。

2. 内容

本书基于企业级大数据采集与预处理项目，以任务驱动为主线，可指导学生进行网络数据的爬取、日志数据的获取，并利用相关大数据预处理技术和工具，根据业务需要进行数据的预处理。本书讲解了平台搭建与运维、大数据采集与存储、大数据预处理等完整的应用案例，全书主要介绍以下技术及应用。

（1）Scrapy 平台的搭建与应用。

（2）Apache Flume 平台的搭建与应用。

（3）爬虫基本库 urllib 和 Selenium 的介绍与使用。

（4）BeautifulSoup 和 PyQuery 解析库的介绍与使用。

（5）Pig 系统环境的搭建与应用。

（6）Kettle 系统环境的搭建与应用。

（7）OpenRefine 平台的搭建与应用。

（8）Pandas 数据分析包的介绍与使用。

3. 使用

（1）教学内容学时安排

本书建议授课 64 学时，教学单元与学时安排如下表所示。

教学单元与学时安排表

序号	课程项目	课程模块（任务、情境）	模块学时	项目学时
1	项目 1　数据采集与预处理准备	任务 1　认识数据采集技术，熟悉数据采集平台	6	12
		任务 2　认识数据预处理技术	6	
2	项目 2　网络爬虫实践	任务 1　使用 urllib 爬取北京公交线路信息	4	16
		任务 2　使用 Selenium 爬取淘宝网站信息	4	
		任务 3　使用 Scrapy 爬取北京公交信息	4	
		任务 4　创新与拓展	4	
3	项目 3　日志数据采集实践	任务 1　Flume 的安装和配置	4	12
		任务 2　Flume 采集数据上传到集群	4	
		任务 3　创新与拓展	4	
4	项目 4　数据预处理实践	任务 1　用 Pig 进行数据预处理	4	24
		任务 2　用 Kettle 进行数据预处理	4	
		任务 3　用 Pandas 进行数据预处理	4	
		任务 4　用 OpenRefine 进行数据预处理	4	
		任务 5　用 Flume Interceptor 对日志信息进行数据预处理	4	
		任务 6　创新与拓展	4	
5	合计			64

（2）课程资源

本书是云计算与大数据等专业校企产教融合系列教材之一，配有丰富的数字化教学资源，包括资源包、运行脚本、电子教案等，读者可以联系北京西普阳光教育科技股份有限公司获得，或登录人邮教育社区（www.ryjiaoyu.com）下载使用。

4．致谢

本书编者包括南京交通职业技术学院的米洪、张鸰、季丹，南京信息职业技术学院的史律，南京工业职业技术学院的张俊，北京西普阳光教育科技股份有限公司的林雪纲，全书由上述老师共同编写并完成项目实践操作验证工作。本书由米洪、张鸰任主编并完成统稿，由季丹、史律、张俊任副主编，林雪纲参与编写。本书的出版得到了南京交通职业技术学院的刘丽奔、尚修利等的大力支持，在此表示衷心的感谢！编者在编写本书过程中参阅了国内外同行编写的相关著作和文献，谨向各位作者致以深深的谢意！

由于编者水平有限，书中难免存在疏漏和不足之处，请广大读者批评、指正。编者联系方式：xxmh@njitt.edu.cn。

编者

2019.7

目录 CONTENTS

项目 1

项目 2

项目3

项目4

项目 1 数据采集与预处理准备

学习目标

【知识目标】

① 识记数据采集与预处理的概念和目的。
② 领会数据采集与预处理的意义。

【技能目标】

① 熟悉数据采集技术。
② 熟悉数据预处理技术。
③ 学会数据采集与预处理环境的搭建。

项目描述

数据采集在大数据处理的生命周期中处于第一个环节。对于不同的数据源，可能存在不同的异构数据集，如文件、XML 树、关系表等，因此需要对这些异构的数据做进一步集成或整合处理，将来自不同数据集的数据进行收集、整理、清洗、转换后，生成一个新的数据集，为后续查询和分析处理提供统一的数据视图。如何进行数据采集，如何对采集的数据进行预处理，去除其中的“脏数据”，是大数据分析面临的问题。

任务 1 认识数据采集技术，熟悉数据采集平台

任务描述

（1）学习数据采集等相关知识内容，熟悉大数据的定义、大数据的基本特征及数据采集的相关技术、工具和产品等。

（2）熟悉数据采集的来源和方法。

（3）完成 Scrapy 平台的搭建。

（4）完成 Apache Flume 平台的搭建。

任务目标

（1）掌握数据采集的来源和采集的方法。

（2）学会搭建数据采集平台。

知识准备

大数据指无法在一定时间范围内用常规工具进行捕捉、管理和处理的数据集合，是需要新处理模式才能具有更强的决策力、洞察发现力和流程优化能力的海量、高增长率和多样化的信息资产。当前，全球大数据进入加速发展时期。大数据时代，谁掌握了足够的数据，谁就有可能掌握未来，现在的数据采集就是将来的资产积累。

1．数据采集的概念

足够的数据量是企业大数据战略建设的基础，因此数据采集成为大数据分析的前站。数据采集是大数据价值挖掘中重要的一环，其后的分析挖掘都建立在数据采集的基础上。大数据技术的意义确实不在于掌握规模庞大的数据信息，而在于对这些数据进行智能处理，从中分析和挖掘出有价值的信息，但前提是拥有大量的数据。

数据的采集有基于物联网传感器的采集，也有基于网络信息的采集。例如，在智能交通中，数据的采集有基于 GPS 的定位信息采集、基于交通摄像头的视频采集、基于交通卡口的图像采集、基于路口的线圈信号采集等。而互联网中的数据采集是对各类网络媒介，如搜索引擎、新闻网站、论坛、微博、博客、电商网站等的各种页面信息和用户访问信息进行采集，采集的内容主要有文本信息、URL、访问日志、日期和图片等。之后需要对采集到的各类数据进行清洗、过滤、去重等预处理并分类归纳存储。

数据采集过程中涉及数据的抽取（Extract）、数据的清洗转换（Transform）、数据的加载（Load）3 个过程，其英文缩写为 ETL。

数据采集的 ETL 工具负责将分布的、异构数据源中的不同种类和结构的数据，如文本数据、关系数据及图片、视频等非结构化数据抽取到临时中间层，然后进行清洗、转换、分类、集成，最后加载到对应的数据存储系统（如数据仓库）中，成为联机分析处理、数据挖掘的基础。

针对大数据的 ETL 处理过程有别于传统的 ETL 处理过程，因为大数据的体量巨大，产生速度也非常快，例如，一个城市的视频监控摄像头、智能电表每一秒都在产生大量的数据，对数据的预处理需要实时快速，因此，在 ETL 的架构和工具选择上，也会采用如分布式内存数据库、实时流处理系统等现代信息技术。

现代企业中存在不同的应用和各种数据格式及存储需求，但在企业之间、企业内部都存在条块分割、信息孤岛的现象，各个企业之间的数据不能实现可控的数据交换和共享，而且各个应用之间开发技术和环境的限制阻碍了企业各个应用之间的数据交换与共享，也增强了企业在数据可控、数据管理、数据安全方面的需求。为实现跨行业、跨部门的数据整合，尤其是在智慧城市建设中，需要制定统一的数据标准、交换接口及共享协议，这样不同行业、不同部门、不同格式的数据才能基于一个统一的基础进行访问、交换和共享。

2. 数据采集的来源

根据 MapReduce 产生数据的应用系统分类，大数据的采集主要有 4 种来源：管理信息系统、Web 信息系统、物理信息系统、科学实验系统。

（1）管理信息系统

管理信息系统是指企业、机关内部的信息系统，如事务处理系统、办公自动化系统，主要用于经营和管理，为特定用户的工作和业务提供支持。数据的产生既有终端用户的始输入，又有系统的二次加工处理。系统的组织结构是专用的，数据通常是结构化的。

（2）Web 信息系统

Web 信息系统包括互联网中的各种信息系统，如社交网站、社会媒体、系统引擎等，主要用于构造虚拟的信息空间，为广大用户提供信息服务和社交服务。系统的组织结构是开放式的，大部分数据是半结构化或无结构的。数据的产生者主要是在线用户。

（3）物理信息系统

物理信息系统是指关于各种物理对象和物理过程的信息系统，如实时监控、实时检测，主要用于生产调度、过程控制、现场指挥、环境保护等。系统的组织结构是封闭的，数据由各种嵌入式传感设备产生，可以是关于物理、化学、生物等性质和状态的基本测量值，也可以是关于行为和状态的音频、视频等多媒体数据。

（4）科学实验系统

科学实验系统实际上也属于物理信息系统，但其实验环境是预先设定的，主要用于学术研究等，数据是有选择的、可控的，有时可能是人工模拟生成的仿真数据。数据往往具有不同的形式。

管理信息系统和 Web 信息系统属于人与计算机的交互系统，物理信息系统属于物与计算机的交互系统。关于物理世界的原始数据，在人与计算机的交互系统中，是通过人实现融合处理的；而在物与计算机的交互系统中，需要通过计算机等装置做专门的处理。融合处理后的数据被转换为规范的数据结构，输入并存储在专门的数据管理系统中，如文件或数据库，形成专门的数据集。

对于不同的数据集，可能存在不同的结构和模式，如文件、XML 树、关系表等，表现为数据的异构性。对于多个异构的数据集，需要做进一步集成处理或整合处理，将来自不同数据集的数据收集、整理、清洗、转换后，生成一个新的数据集，为后续查询和分析处理提供统一的数据视图。

3. 数据采集的方法

（1）数据采集的新方法

① 系统日志采集方法

很多互联网企业有自己的海量数据采集工具，多用于系统日志采集，如 Hadoop 的 Chukwa、Cloudera 的 Flume、Facebook 的 Scribe 等，这些工具均采用分布式架构，能满足每秒数百兆字节的日志数据的采集和传输需求。

② 网络数据采集方法：对非结构化数据的采集

网络数据采集是指通过网络爬虫或网站公开 API 等方式从网站上获取数据信息，该方法可以将非结构化数据从网页中抽取出来，将其存储为统一的本地数据文件，并以结构化的方式存储。网络数据采集方法支持图片、音频、视频等文件或附件的采集，附件与正文可以自动关联。

除了网络中包含的内容之外，对于网络流量的采集可以使用 DPI 或 DFI 等带宽管理技术进行处理。

③ 其他数据采集方法

对于企业生产经营数据或学科研究数据等对保密性要求较高的数据，可以通过与企业、研究机构合作或授权的方式，使用特定系统接口等采集数据。

（2）网页数据采集的方法

互联网网页数据具有分布广、格式多样、非结构化等大数据的典型特点，需要有针对性地对互联网网页数据进行采集、转换、加工和存储。在网页数据的架构和处理方面，存在急需突破的若干关键技术。

传统的数据挖掘、分析处理方法和工具，在非结构化、高速化的大数据处理要求面前显得过于乏力，需要创新开发适应新型大数据处理需求的数据挖掘和数据处理方法。

互联网网页数据是大数据领域的一个重要组成部分，是互联网公司和金融机构获取用户消费、交易、产品评价信息以及其他社交信息等数据的重要途径，为互联网和金融服务创新提供了丰富的数据基础，因此，对互联网网页的大数据处理流程和技术进行探索具有重要意义。

① 网页数据采集的基本流程

互联网网页数据采集就是获取互联网中相关网页内容的过程，并从中抽取出用户所需要的属性内容。互联网网页数据处理，就是对抽取出来的网页数据进行内容和格式上的处理，进行转换和加工，使之能够适应用户的需求，并将之存储下来，以供后用。

网络爬虫是一个自动提取网页的程序，它为搜索引擎从万维网上下载网页，是搜索引擎的重要组成部分。传统爬虫从一个或若干初始网页的 URL 开始，获得初始网页的 URL，在抓取网页的过程中，不断从当前页面中抽取新的 URL 放入队列，直到满足系统的一定停止条件。

聚焦爬虫的工作流程较为复杂，需要根据一定的网页分析算法过滤与主题无关的链接，保留有用的链接并将其放入等待抓取的 URL 队列。它将根据一定的搜索策略从队列中选择下一步要抓取的网页 URL，并重复上述过程，直到达到系统的某一条件时停止。

另外，所有被爬虫抓取的网页将会被系统存储起来，进行一定的分析、过滤，并建立索引，以便之后的查询和检索；对于聚焦爬虫来说，这一过程所得到的分析结果还可能对以后的抓取过程给出反馈和指导。网络爬虫自动提取网页的过程如图 1-1 所示。

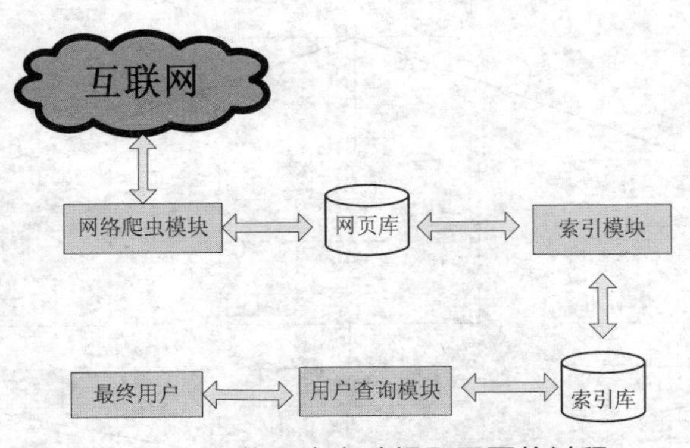

图 1-1 网络爬虫自动提取网页的过程

② 网页数据采集的工作过程

数据采集的目的就是把目标网站上网页中的某块文字或者图片等资源下载到指定位置。这个过程需要做如下配置工作：下载网页配置，解析网页配置，修正结果配置，数据输出配置。如果数据符合自己的要求，则修正结果这一步可省略。配置完毕后，把配置形成任务（任务以 XML 格式描述），采集系统按照任务的描述开始工作，最终把采集到的结果存储到指定位置。

整个数据采集过程的基本步骤如下。

a. 将需要抓取数据网站的 URL（Site URL）信息写入 URL 队列。

b. 爬虫从 URL 队列中获取需要抓取数据网站的 URL 信息。

c. 获取某个具体网站的网页内容。

d. 从网页内容中抽取出该网站正文页内容的链接地址。

e. 从数据库中读取已经抓取过内容的网页地址。

f. 过滤 URL。对当前的 URL 和已经抓取过的 URL 进行比较。

g. 如果该网页地址没有被抓取过，则将该地址写入数据库。如果该地址已经被抓取过，则放弃对这个地址的抓取操作。

h. 获取该地址的网页内容，并抽取出所需属性的内容值。

i. 将抽取的网页内容写入数据库。

数据采集工作的流程如图 1-2 所示。

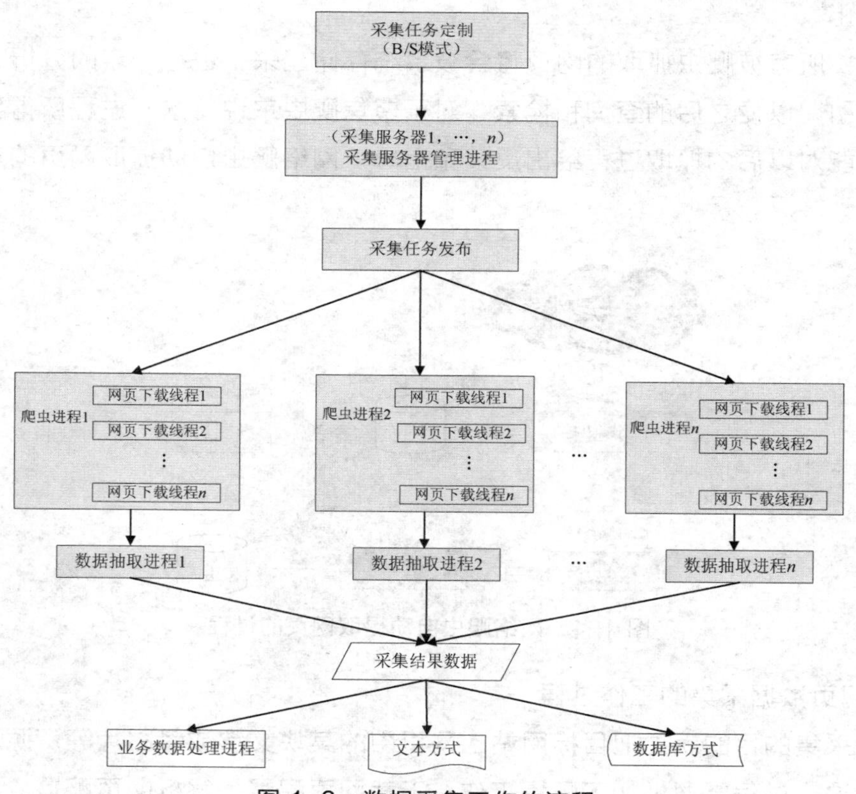

图 1-2　数据采集工作的流程

相应的网页内容提取、数据采集与数据处理逻辑如图 1-3 所示。

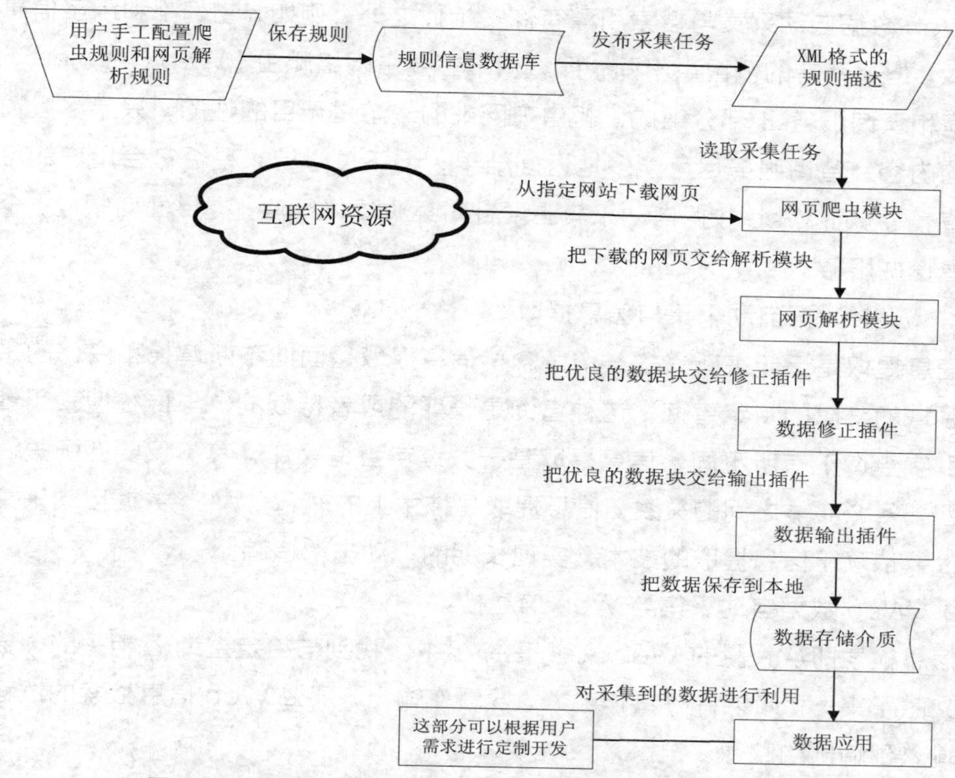

图 1-3　相应的网页内容提取、数据采集与数据处理逻辑

③ Web 信息数据自动采集

Web 可以说是目前最大的信息系统，其数据具有海量、多样、异构、动态变化等特性。因此，人们要准确迅速地获得自己所需要的数据越来越难，尽管目前有各种搜索引擎，但是搜索引擎在数据的查全率方面考虑较多，而查准率不足，而且很难进一步挖掘深度数据。所以，人们开始研究如何更进一步地获取互联网中某一个特定范围内的数据，从信息搜索到知识发现。

Web 数据自动采集涉及 Web 数据挖掘（Web Data Mining）、信息检索（Information Retrieval）、信息提取（Information Extraction）、搜索引擎（Search Engine）等概念和技术。这些概念密切相关，但又有所区别。

a. Web 数据自动采集与挖掘

所谓 Web 数据自动采集与挖掘，是指从大量非结构化、异构的 Web 信息资源中发现有效的、新颖的、潜在可用的及最终可以理解的知识（包括概念、模式、规则、约束及可视化等形式）的过程。

b. Web 数据自动采集与搜索引擎

Web 数据自动采集与搜索引擎有许多相似之处，例如，它们都利用了信息检索的技术。但是两者的侧重点不同，搜索引擎主要由网络爬虫（Web Crawler）、索引数据库和查询服务 3 部分组成。爬虫在网上的漫游是无目的性的，只是尽量发现比较多的内容。查询服务尽可能多地返回结果，但不关心结果是否符合用户的习惯、专业背景等。而 Web 数据自动采集主要针对某个具体行业，提供面向领域、个性化的信息挖掘服务。

c. Web 数据自动采集与信息提取

信息提取是近年来新兴的一个概念。信息提取是面向不断增长和变化的、某个具体领域的文献的特定查询，这种查询是长期的或者持续的。与传统搜索引擎（基于关键字查询）有所不同，信息提取基于关键字查询操作，其不仅要匹配关键字，还要匹配各个实体之间的关系。信息提取是技术上的概念。Web 数据自动采集很大程度上要依赖于信息提取的技术，实现长期的、动态的追踪。

d. Web 数据自动采集与 Web 信息检索

信息检索即从大量的 Web 文献集合 C 中，找到与给定查询 q 相关的、数目相当的文献子集 S。如果将 q 看作输入，S 看作输出，那么 Web 信息检索的过程就是一个输入到输出的映像：

$$\S: (C: q) \to S$$

而 Web 数据自动采集不是将 Web 文献集合的子集直接输出给用户，而是还要进一步分析处理，如查重、去噪、整合数据等。尽量将半结构化甚至非结构化的数据变为结构化的数据，并以统一的格式呈现给用户。

因此，Web 数据自动采集是 Web 数据挖掘的一个重要组成部分，它利用了 Web 数据检索、信息提取的技术，弥补了搜索引擎缺乏针对性和专业性，无法实现数据的动态跟踪与监测的缺点，是一个非常有前景的领域。

④ 链接过滤

链接过滤是数据采集的关键技术，实质就是判断一个链接（当前链接）是不是在一个链接集合（已经抓取过的链接）中。在对网页大数据的采集中，可以采用布隆过滤器（Bloom Filter）来实现对链接的过滤。

布隆过滤器的基本思想是，当一个元素被加入集合时，通过 k 个散列函数将这个元素映射成一个位数组中的 k 个点，把它们置为1。检索时，只要看看这些点是不是都是1 就知道集合中有没有被检元素：如果这些点有任何一个是 0，则被检元素一定不在；如果都是 1，则被检元素很可能存在。

布隆过滤器的具体实现方法是：已经抓取过的每个 URL，经过 k 个 Hash 函数的计算，得出 k 个值，再和一个巨大的位数组中的这 k 个位置的元素对应起来（这些位置数组元素的值被设置为 1）。在需要判断某个 URL 是否被抓取过时，先用 k 个 Hash 函数对该 URL 计算出 k 个值，再查询巨大的位数组内这 k 个位置上的值，如果全为 1，则已经被抓取过，否则没有被抓取过。

⑤ Web 引擎和通用搜索引擎的差别

Web 引擎和通用搜索引擎相比有较大的差别，Web 引擎更多地关注“结构化信息”的抽取。如比较购物搜索时需要在抓取网页内容后，进一步抽取出商品名称、价格、服务、简介等；在房产信息搜索中应抽取出类型、地域、地址、房型、面积、装修情况、租金、联系人、联系电话等。

通用搜索引擎是指从互联网检索出满足搜索条件的信息反馈给用户。更多关注搜索条件，信息一般不进行结构化处理。

⑥ 结构化信息抽取的方式

Web结构化信息抽取就是将网页中的非结构化数据按照一定的需求抽取成结构化数据，属于垂直搜索。

结构化信息抽取有两种方式可以实现，比较简单的是模板方式，另一种是对网页不依赖的网页库级的结构化信息抽取方式。

a. 模板方式

模板方式是事先对特定的网页进行模板配置，抽取模板中设置好的需要的信息，可以针对有限个网站的信息进行精确采集。

特点：简单、精确、技术难度低、方便快速部署。

缺点：需要针对每一个信息源的网站模板进行单独设定，在信息源多样性强的情况下维护量巨大，甚至是不可完成的维护量。所以这种方式适用于少量信息源的信息处理，其不能作为搜索引擎级的应用，很难满足用户对查全率的需求。

b. 网页库级方式

这种方式采用页面结构分析与智能结点分析转换的方法，自动抽取结构化的数据。

特点：可对任意的正常网页进行抽取，完全自动化，不用对具体网站事先生成模板，对每个网页自动实时地生成抽取规则，完全不需要人工干预；智能抽取准确率高，不是机械的匹配，采用智能分析技术，准确率能达到 98%以上；能保证较快的处理速度，由于采用页面的智能分析技术，先去除了垃圾块，降低了分析的压力，因此处理速度大大提高了；通用性较好，易于维护，只需设定参数、配置相应的特征就能改进相应的抽取性能。

缺点：技术难度高，前期研发成本高，周期长，适合网页库级别结构化数据采集和搜索的高端应用。

任务实施

1. Scrapy 系统环境搭建

Scrapy 是 Python 开发的一个快速、高层次的屏幕抓取和 Web 抓取框架，任何人都可以根据需求方便地对它进行修改，用于从 Web 页面中提取结构化的数据，Scrapy 提供了多种类型爬虫的基类，支持 BaseSpider、Sitemap、Web2.0 等爬虫。

（1）安装所需的环境

右键单击 Ubuntu 操作系统桌面，在弹出的快捷菜单中选择“Open Terminal”命令，打开命令行窗口，在其中输入命令“sudo apt-get install python-pip”，安装 pip，如图 1-4 所示。

```
li@li-virtual-machine:~$ sudo apt-get install python-pip
```

图 1-4　使用命令安装 pip

pip 是一个现代的、通用的 Python 包管理工具，提供了对 Python 包的查找、下载、安装、卸载的功能。需要将 pip 更新到最新版本，即在命令行窗口中执行命令“pip install --upgrade pip”，如图 1-5 所示。

```
li@li-virtual-machine:~$ pip install --upgrade pip
```

图 1-5　更新 pip 到最新版本

pip 更新完成后，即可安装 Scrapy。在命令行窗口中执行命令“pip install scrapy”，如图 1-6 所示。

```
li@li-virtual-machine:~$ pip install scrapy
```

图 1-6　安装 Scrapy

（2）验证 Scrapy 框架安装是否成功

在命令行窗口中输入命令“scrapy”，若显示图 1-7 所示的内容，即表示成功安装 Scrapy 框架。

```
li@li-virtual-machine:~$ scrapy
Scrapy 1.5.1 - no active project

Usage:
  scrapy <command> [options] [args]

Available commands:
  bench         Run quick benchmark test
  fetch         Fetch a URL using the Scrapy downloader
  genspider     Generate new spider using pre-defined templates
  runspider     Run a self-contained spider (without creating a project)
  settings      Get settings values
  shell         Interactive scraping console
  startproject  Create new project
  version       Print Scrapy version
  view          Open URL in browser, as seen by Scrapy

  [ more ]      More commands available when run from project directory

Use "scrapy <command> -h" to see more info about a command
```

图 1-7　验证 Scrapy 框架安装是否成功

2. 日志系统环境搭建

Flume 是 Cloudera 提供的一个高可用的、高可靠的、分布式的海量日志采集、聚合和传输系统，Flume 支持在日志系统中定制各类数据发送方，用于收集数据；同时，Flume 具有对数据进行简单处理，并写到各种数据接收方（可定制）的能力。

（1）安装 Flume

Flume 需要 JDK 环境的支持，可以使用“java –version”命令查看系统是否配置了 JDK 环境，若显示如图 1-8 所示的结果，即表示配置了 JDK 环境。

```
a@a:/usr/local/flume$ java -version
java version "1.8.0_181"
Java(TM) SE Runtime Environment (build 1.8.0_181-b13)
Java HotSpot(TM) 64-Bit Server VM (build 25.181-b13, mixed mode)
a@a:/usr/local/flume$
```

图 1-8　验证是否配置了 JDK 环境

否则，需要先下载并安装 JDK 环境。

① 到官方网站下载

执行以下命令，完成压缩包的解压与安装。

```
cd ~/Downloads
sudo tar -zxvf apache-flume-1.8.0-bin.tar.gz -C /usr/local
```

② 修改权限和命名

首先，使用“id”命令确定本机的用户和组，如图 1-9 所示。

```
a@a:/usr/local$ id
uid=1000(bailing) gid=1000(bailing) groups=1000(bailing),4(a
dm),24(cdrom),27(sudo),30(dip),46(plugdev),113(lpadmin),128(
sambashare)
```

图 1-9　确定本机的用户和组

其次，使用以下命令更改用户和组修改文件的权限，如图 1-10 所示。

```
cd /usr/local
sudo chown -R a(用户):a(组) apache-flume-1.7.0-bin
```

```
a@a:/usr/local$ sudo chown -R a:a apache-flume-1.8.0-bin
a@a:/usr/local$ 
```

图 1-10　更改用户和组修改文件的权限

修改文件名称，如图 1-11 所示。

```
a@a:/usr/local$ sudo mv apache-flume-1.8.0-bin flume
a@a:/usr/local$ 
```

图 1-11　修改文件名称

（2）配置环境变量

① 配置环境变量

执行“sudo gedit /etc/profile”命令，在 profile 文件中配置环境变量，如图 1-12 所示。

```
export FLUME_HOME=/usr/local/flume
export FLUME_CONF_DIR=$FLUME_HOME/conf
export PATH=$JAVA_HOME/bin:$JRE_HOME/bin:$PATH:$FLUME_HOME/bin
```

图 1-12　配置环境变量

配置完成并保存后，需要使用以下命令使环境变量生效。

```
Source /etc/profile
```

② 修改配置文件 flume-env.sh

进入/usr/local/flume/conf 目录，进行图 1-13 所示的操作。

```
a@a:/usr/local/flume/conf$ sudo mv flume-env.sh.template flume-env.sh
a@a:/usr/local/flume/conf$ 
```

图 1-13　文件重命名

在 flume-env.sh 文件开头加入图 1-14 所示的语句。

```
export JAVA_HOME=/usr/local/java/jdk1.8.0_181
```

图 1-14　要加入的语句

（3）验证 Flume 框架安装是否成功

执行以下语句。

```
cd   /usr/local/flume
./bin/flume-ng version
```

若显示图 1-15 所示的信息，则表示 Flume 安装成功。

```
a@a:/usr/local$ ./flume/bin/flume-ng version
Flume 1.8.0
Source code repository: https://git-wip-us.apache.org/repos/
asf/flume.git
Revision: 99f591994468633fc6f8701c5fc53e0214b6da4f
Compiled by denes on Fri Sep 15 14:58:00 CEST 2017
From source with checksum fbb44c8c8fb63a49be0a59e27316833d
```

图 1-15　验证 Flume 是否安装成功

任务 2　认识数据预处理技术

任务描述

（1）学习数据预处理技术等相关知识内容，如数据清洗的主要任务和常用方法，数据集成的主要任务和常用方法，数据转换的主要任务和常用方法，数据归约的主要任务和常用方法。

（2）完成 Pig 系统环境的搭建。

（3）完成 Kettle 系统环境的搭建。

任务目标

（1）了解原始数据存在的主要问题。

（2）明白数据预处理的作用和工作任务。

（3）学会数据处理工具平台的搭建。

知识准备

1. 数据预处理的概念

数据预处理是指在对数据进行数据挖掘的主要处理以前，先对原始数据进行必要的清理、集成、转换、离散、归约、特征选择和提取等一系列处理工作，达到挖掘算法进行知识获取、研究所要求的最低规范和标准。

数据挖掘的对象是从现实世界采集到的大量的、各种各样的数据。现实生产和实际生活以及科学研究的多样性、不确定性、复杂性等导致采集到的原始数据比较散乱，它们是不符合挖掘算法进行知识获取、研究所要求的规范和标准的，这些数据主要具有以下特征。

（1）不完整性。不完整性指的是数据记录中可能会出现有一些数据属性的值丢失或不确定的情况，还有可能缺失必需的数据。这是系统设计时存在的缺陷或者使用过程中一些人为因素造成的，如有些数据缺失只是因为输入时被认为是不重要的，相关数据没有记录可能是由于理解错误，或者因为设备故障，与其他记录不一致的数据可能已经被删除，历史记录或修改的数据可能被忽略等。

（2）含噪声。含噪声指的是数据具有不正确的属性值，包含错误或存在偏离期望的离群值（指与其他数值比较差异较大的值）。它们产生的原因有很多，如收集数据的设备可能出现了故障，人或计算机可能在数据输入时出现了错误，数据传输中可能出现了错误等。不正确的数据也可能是由命名约定或所用的数据代码不一致，或输入字段（如时间）的格式不一致而导致的。在实际使用的系统中，还可能存在大量的模糊信息，有些数据甚至具有一定的随机性。

（3）杂乱性（不一致性）。原始数据是从各个实际应用系统中获取的，由于各应用系统的数据缺乏统一标准的定义，数据结构也有较大的差异，因此各系统间的数据存在较大的不一致性，往往不能直接使用。同时，来自不同应用系统中的数据，由于合并还普遍存在数据重复和信息冗余现象。

因此，这里说存在不完整的、含噪声的和不一致的数据是现实世界大型的数据库或数据仓库的共同特点。一些比较成熟的算法对其处理的数据集合一般有一定的要求，如数据完整性好、数据的冗余性小、属性之间的相关性小。然而，实际系统中的数据一般无法直接满足数据挖掘算法的要求，因此必须对数据进行预处理，以提高数据质量，使之符合数据挖掘算法的规范和要求。

2. 数据预处理的常见问题

（1）数据采样

数据采样技术分为加权采样、随机采样和分层采样 3 类，其目的是从数据集中采集部分样本进行处理。

加权采样的思想是通过对总体中的各个样本设置不同的数值系数（即权重），使样本呈现希望的相对重要性程度。

随机采样是最常用的方法。许多算法在初始化时会计算数据的随机样本，随机样本可以利用事先准备好的已经排序的随机数表得到。但是，有时为了得到更高的性能，希望能够随时取得随机样本，而使用随机函数可以达成这个目的。

分层采样的思想是根据数据分布的不均衡性控制采样的频率。在数据分布密度较高时，采样的频率应适当降低；在数据分布密度较低时，采样的频率应适当提高。

（2）数据清理

数据清理技术通常包括填补遗漏的数据值、平滑有噪声数据、识别或除去异常值，以及解决不一致问题。填补遗漏的数据值，处理不完备数据集的方法主要有以下三大类。

① 删除元组

删除元组就是将存在遗漏信息属性值的对象（元组，记录）删除，从而得到一个完备的信息表。这种方法简单易行，在对象有多个属性缺失值、被删除的含缺失值的对象数量与信息表中的数据量相比非常小的情况下是非常有效的。然而，这种方法有很大的局限性。它减少历史数据来换取信息的完备，这会造成资源的大量浪费，丢弃大量隐藏在这些对象中的信息。在信息表中本来包含对象很少的情况下，删除少量对象就足以严重影响到信息表信息的客观性和结果的正确性；当每个属性空值的百分比变化很大时，它的性能会非常差。因此，当遗漏数据所占比例较大，特别是当遗漏数据非随机分布时，这种方法可能导致数据发生偏离，从而引出错误的结论。

② 数据补齐

这类方法是用一定的值去填充空值，从而使信息表完备化。通常基于统计学原理，根据决策表中其余对象取值的分布情况来对一个空值进行填充，譬如用其余属性的平均值来进行补充等。数据挖掘中常用的补齐方法有以下几种。

a. 人工填写：由于最了解数据的是用户自己，因此这个方法产生的数据偏离最小，可能是填充效果最好的一种。然而，一般来说，此方法很费时，当数据规模很大、空值很多的时候，此方法是不可行的。

b. 特殊值填充：将空值作为一种特殊的属性值来处理，它不同于其他的任何属性值，如所有的空值都用“unknown”填充。如果该缺失值与预测值具有高度相关性，可能导致严重的数据偏离，一般不推荐使用。

c. 平均值填充：将信息表中的属性分为数值属性和非数值属性来分别进行处理。如果空值是数值型的，则使用该属性在其他所有对象的取值的平均值来填充该缺失的属性值；如果空值是非数值型的，则根据统计学中的众数原理，用该属性在其他所有对象的取值次数最多的值（即出现频率最高的值）来补齐该缺失的属性值。另外，有一种与其相似的方法称为条件平均值填充法。在此方法中，缺失属性值的补齐同样是靠该属性在其他对象中的取值求平均得到的，但不同的是，用于求平均的值并不是从信息表所有对象中取得的，而是从与该对象具有相同决策属性值的对象中取得的。这两种数据的补齐方法，其基本的出发点都是一样的，以最大概率可能的取值来补充缺失的属性值，只是在具体方法上有一点不同。与其他方法相比，它们用现存数据的多数信息来推测缺失值。

d．热卡填充（或就近补齐）：对于一个包含空值的对象，热卡填充法在完整数据中找到一个与它最相似的对象，并用这个相似对象的值来进行填充。不同的问题可能会选用不同的标准来对相似进行判定。此方法在概念上很简单，且利用了数据间的关系来进行空值估计。此方法的缺点在于难以定义相似标准，主观因素较多。

e．k近邻法：先根据欧式距离或相关分析来确定距离具有缺失数据样本最近的k个样本，再将这k个值加权平均来估计该样本的缺失数据。

f．使用所有可能的值填充：这种方法用空缺属性值的所有可能的属性取值来填充，能够得到较好的补齐效果。但是，当数据量很大或者遗漏的属性值较多时，其计算的代价很大，可能的测试方案很多。还有一种方法，其填补遗漏属性值的原则是一样的，不同的只是其从决策相同的对象中尝试所有的属性值的可能情况，而不是根据信息表中的所有对象进行尝试，这样能够在一定程度上减小原方法的代价。

g．组合完整化方法：这种方法用空缺属性值的所有可能的属性取值来尝试，并从最终属性的约简结果中选择最好的一个作为填补的属性值。这是以约简为目的的数据补齐方法，能够得到好的约简结果；但是，当数据量很大或者遗漏的属性值较多时，其计算的代价很大。另一种方法称为条件组合完整化方法，其填补遗漏属性值的原则是一样的，不同的只是其从决策相同的对象中尝试所有的属性值的可能情况，而不是根据信息表中所有对象进行尝试。条件组合完整化方法能够在一定程度上减小组合完整化方法的代价。在信息表包含不完整数据较多的情况下，可能的测试方案将剧增。

h．回归：基于完整的数据集，建立回归方程（模型）。对于包含空值的对象，将已知属性值代入方程可估计未知属性值，以此估计值来进行填充。当变量不是线性相关或是与预测变量高度相关时，会导致有偏差的估计。

③ 平滑有噪声数据

噪声是一个测量变量中的随机错误或偏差，包含错误值或偏离期望的孤立点值。数据平滑技术包括以下几种。

a．分箱：分箱方法通过考察数据的“近邻”（即周围的值）来平滑有序数据值。这些有序的值被分布到一些桶或箱中。由于分箱方法考察邻近的值，因此它进行的是局部平滑。如果使用箱均值平滑，则箱中每一个值被箱中的平均值替换；如果使用箱中位数平滑，则箱中的每一个值被箱中的中位数替换；如果使用箱边界平滑，则箱中的最大值和最小值同样被视为边界，箱中的每一个值被最近的边界值替换。

一般而言，宽度越大，平滑效果越明显。箱也可以是等宽的，其中每个箱值的区间范围是一个常量。分箱也可以作为一种离散化技术使用。

b. 回归：也可以用一个函数拟合数据来平滑数据。线性回归涉及找出拟合两个属性（或变量）的“最佳”直线，使得一个属性能够预测另一个属性。多线性回归是线性回归的扩展，它涉及两个以上的属性，并且数据会拟合到一个多维面。使用回归，找出适合数据的数学方程式，能够帮助消除噪声。

c. 聚类：可以通过聚类来检测离群点，将类似的值组织成群或簇。直观地讲，落在簇集合之外的值被视为离群点。

（3）数据集成

数据集成指将来自多个数据源的数据合并，形成一致的数据存储，如将不同数据库中的数据集成到一个数据仓库中存储。有时数据集成之后还需要进行数据清理，以便消除可能存在的数据冗余。在数据集成时需要考虑很多问题。

① 实体识别问题

模式集成和对象匹配可能需要技巧。来自多个信息源的现实世界的等价实体如何才能“匹配”？这涉及实体识别问题。例如，数据分析者或计算机如何才能确信一个数据库中的cust_id和另一个数据库中的cust_number指的是同一实体？每个属性的元数据包括名称、含义、数据类型和属性的允许取值范围，以及处理空白、零或 NULL 值的空值规则。这种元数据可以帮助用户避免模式集成的错误。元数据还可以用来帮助用户转换数据。

在集成期间，当一个数据库的属性与另一个数据库的属性匹配时，必须特别注意数据的结构。这旨在确保源系统中的函数依赖和参照约束与目标系统中的匹配。

② 冗余和相关分析

冗余是数据集成的另一个重要问题。一个属性（如年收入）如果能由另一个或另一组属性“导出”，则这个属性可能是冗余的。属性命名的不一致也可能导致数据集成过程中的冗余。

③ 元组重复

除了检测属性间的冗余之外，还应当在元组级检测重复。例如，对于给定的唯一数据实体，存在两个或多个相同的元组。

④ 数据值冲突的检测与处理

数据集成还涉及数据值冲突的检测与处理。例如，对于现实世界的同一实体，来自不同数据源的属性值可能不同。这可能是因为表示、尺度或编码不同。例如，重量属性可能在一个系统中以公制单位存放，而在另一个系统中以英制单位存放。

属性也可能在不同的抽象层中，其中属性在一个系统中记录的抽象层可能与另一个系统中相同名称属性记录的具有不同含义。

⑤ 数据转换

数据转换主要是指将数据转换成适用于挖掘的形式，如将属性数据按比例缩放，使之落入一个比较小的特定区间，这一点对那些基于距离的挖掘算法尤为重要。数据转换的具体方法包括平滑处理、聚集处理、数据泛化处理、规格化、属性构造。

⑥ 数据归约

数据归约指在不影响挖掘结果的前提下，通过数值聚集、删除冗余特性的办法压缩数据，提高挖掘模式的质量，降低时间复杂度。

数据归约策略包括维归约、数量归约和数据压缩。

维归约减少了所考虑的随机变量或属性的个数。用于分析的数据集可能包含数以百计的属性，其中大部分属性与挖掘任务不相关或冗余。例如，分析银行客户的信用度时，诸如客户的电话号码、家庭住址等属性就与该数据挖掘任务不相关，或者说是冗余的。维归约通过减少不相关的属性（或维）达到减小数据集规模的目的。

数量归约用替代的、较小的数据表示形式替换原数据。

数据压缩使用转换，以便得到原数据的归约或“压缩”表示。如果原数据能够从压缩后的数据重构而不损失信息，则该数据归约为无损的。如果只能近似重构原数据，则该数据归约为有损的。

⑦ 特征选择

将高维空间的样本通过映射或者变换的方式转换到低维空间，可达到降维的目的，并通过特征选取删除冗余和不相关的特征来进一步降维。特征选择是从原始特征中挑选出一些最优代表性的特征，它分为过滤式、封装式和嵌入式 3 种类型。

过滤式的主要思想是，对每一维的特征“打分”，即给每一维的特征赋予权重，这样的权重就代表该维特征的重要性，并依据权重排序。其主要方法有 chi-squared test（卡方检验）、information gain（信息增益）、correlation coefficient scores（相关系数）。

封装式的主要思想是，将子集的选择看作一个搜索寻优问题，生成不同的组合，对组合进行评价，再与其他的组合进行比较。这样即可将子集的选择看作一个优化问题，有很多的优化算法可以解决这个问题，尤其是一些启发式的优化算法，如人工蜂群算法（ABC）、粒子群算法（PSO）等。此外也可以采用递归特征消除算法得以实现。

嵌入式的主要思想是，在模型既定的情况下学习对提高模型准确性最好的属性，简单来讲就是在确定模型的过程中，挑选出那些对模型的训练有重要意义的属性。简单易学的机器学习算法——岭回归就是在基本线性回归的过程中加入了正则项。

⑧ 特征提取

特征提取就是利用已有特征参数构造一个较低维数的特征空间，将原始特征中蕴含的有用信息映射到少数几个特征上，忽略多余的不相干信息。简单来说，特征提取是用映射（或转换）的方法把原始特征转换为较少的新特征。特征提取的方法有很多种，如传统的特征提取的数据挖掘技术、统计特征提取技术、神经网络法等。

任务实施

1. Pig 系统环境的搭建

（1）下载 Pig

在官方网站下载 pig-0.17.0-src.tar.gz，并解压到/usr/local 目录，解压操作如图 1-16 所示。

```
hadoop@ubuntu:~/Downloads$ ls
pig-0.17.0.tar.gz
hadoop@ubuntu:~/Downloads$ sudo tar -zxvf pig-0.17.0.tar.gz -C /usr/local
```

图 1-16 解压 Pig 文件到相应目录

解压完成后进入/usr/local，将文件“pig-0.17.0-src”重命名为“pig”，以方便后续使用，如图 1-17 所示。

（2）配置环境变量

打开命令行窗口，输入“sudo vim ~/.bashrc”，配置环境变量，如图 1-18 所示。

```
hadoop@ubuntu: /usr/local
hadoop@ubuntu:/usr/local$ sudo mv pig-0.17.0-src/ pig
hadoop@ubuntu:/usr/local$ ls
bin                                                                   pycharm
etc        lib                                                        sbin
flume      man                                                        share
games      mongodb                                                    src
           myeclipse-2017-ci-10-offline-installer-linux.zip.qbl       test
include    pdi-ce-7.0.0.0-25.zip                                      testflume
java       pig                                                        tomcat
hadoop@ubuntu:/usr/local$
```

图 1-17　对文件进行重命名

```
# 配置Java环境
export JAVA_HOME=/usr/local/java/jdk1.8.0_171
export JRE_HOME=${JAVA_HOME}/jre
export CLASSPATH=.:${JAVA_HOME}/lib:${JRE_HOME}/lib
export PATH=${JAVA_HOME}/bin:$PATH

# 配置Hadoop环境
export HADOOP_HOME=/usr/local/hadoop
export HADOOP_COMMON_LIB_NATIVE_DIR=$HADOOP_HOME/lib/native
export HADOOP_OPTS="-Djava.library.path=$HADOOP_HOME/lib:$HADOOP_COMMON_LIB_NATI
VE_DIR"

# 配置Pig环境
export PIG_INSTALL=/usr/local/pig
export PATH=$PATH:$PIG_INSTALL/bin
                                                        132,26         Bot
```

图 1-18　配置环境变量

环境变量配置完成并保存后，执行“source ～/.bashrc”命令，使配置的环境变量生效。

（3）验证 Pig 是否安装成功

打开两个命令行窗口，分别输入“pig -x local”和“pig -x mapreduce”，测试这两种运行模式，若显示信息分别如图 1-19 和图 1-20 所示，则表示安装成功。

```
hadoop@ubuntu: ~
hadoop@ubuntu:~$ pig -x local
18/07/18 14:53:41 INFO pig.ExecTypeProvider: Trying ExecType : LOCAL
18/07/18 14:53:41 INFO pig.ExecTypeProvider: Picked LOCAL as the ExecType
2018-07-18 14:53:41,301 [main] INFO  org.apache.pig.Main - Apache Pig version 0.
16.0 (r1746530) compiled Jun 01 2016, 23:10:49
2018-07-18 14:53:41,302 [main] INFO  org.apache.pig.Main - Logging error message
s to: /home/hadoop/pig_1531896821299.log
2018-07-18 14:53:41,328 [main] INFO  org.apache.pig.impl.util.Utils - Default bo
otup file /home/hadoop/.pigbootup not found
2018-07-18 14:53:41,697 [main] INFO  org.apache.hadoop.conf.Configuration.deprec
ation - mapred.job.tracker is deprecated. Instead, use mapreduce.jobtracker.addr
ess
2018-07-18 14:53:41,697 [main] INFO  org.apache.hadoop.conf.Configuration.deprec
ation - fs.default.name is deprecated. Instead, use fs.defaultFS
2018-07-18 14:53:41,701 [main] INFO  org.apache.pig.backend.hadoop.executionengi
ne.HExecutionEngine - Connecting to hadoop file system at: file:///
2018-07-18 14:53:42,050 [main] INFO  org.apache.hadoop.conf.Configuration.deprec
ation - io.bytes.per.checksum is deprecated. Instead, use dfs.bytes-per-checksum
2018-07-18 14:53:42,093 [main] INFO  org.apache.pig.PigServer - Pig Script ID fo
r the session: PIG-default-db118730-b7d8-424a-a7cc-df193350b49e
2018-07-18 14:53:42,093 [main] WARN  org.apache.pig.PigServer - ATS is disabled
since yarn.timeline-service.enabled set to false
grunt>
```

图 1-19　Pig 的 Local 模式显示的信息

```
hadoop@ubuntu: ~
grunt> 2018-07-18 14:54:43,473 [main] INFO  org.apache.pig.Main - Pig script com
pleted in 1 minute, 2 seconds and 918 milliseconds (62918 ms)
hadoop@ubuntu:~$ pig -x mapreduce
18/07/18 14:54:51 INFO pig.ExecTypeProvider: Trying ExecType : LOCAL
18/07/18 14:54:51 INFO pig.ExecTypeProvider: Trying ExecType : MAPREDUCE
18/07/18 14:54:51 INFO pig.ExecTypeProvider: Picked MAPREDUCE as the ExecType
2018-07-18 14:54:51,240 [main] INFO  org.apache.pig.Main - Apache Pig version 0.
16.0 (r1746530) compiled Jun 01 2016, 23:10:49
2018-07-18 14:54:51,240 [main] INFO  org.apache.pig.Main - Logging error message
s to: /home/hadoop/pig_1531896891238.log
2018-07-18 14:54:51,270 [main] INFO  org.apache.pig.impl.util.Utils - Default bo
otup file /home/hadoop/.pigbootup not found
2018-07-18 14:54:52,196 [main] INFO  org.apache.hadoop.conf.Configuration.deprec
ation - mapred.job.tracker is deprecated. Instead, use mapreduce.jobtracker.addr
ess
2018-07-18 14:54:52,196 [main] INFO  org.apache.hadoop.conf.Configuration.deprec
ation - fs.default.name is deprecated. Instead, use fs.defaultFS
2018-07-18 14:54:52,196 [main] INFO  org.apache.pig.backend.hadoop.executionengi
ne.HExecutionEngine - Connecting to hadoop file system at: hdfs://localhost:9000
2018-07-18 14:54:53,775 [main] INFO  org.apache.pig.PigServer - Pig Script ID fo
r the session: PIG-default-2d0edc64-a94f-47b8-a93e-fa75b0c214d0
2018-07-18 14:54:53,775 [main] WARN  org.apache.pig.PigServer - ATS is disabled
since yarn.timeline-service.enabled set to false
grunt>
```

图 1-20　Pig 的 MapReduce 模式显示的信息

2. Kettle 系统环境的搭建

（1）下载 Kettle

在官方网站下载 pdi-ce-7.0.0.0-25.zip，并解压到/usr/local 目录，解压操作如图 1-21 所示。

```
hadoop@ubuntu: ~/Downloads
hadoop@ubuntu:~/Downloads$ ls
pdi-ce-7.0.0.0-25.zip  pig-0.17.0-src.tar.gz
hadoop@ubuntu:~/Downloads$ sudo unzip pdi-ce-7.0.0.0-25.zip -d /usr/local
Archive:  pdi-ce-7.0.0.0-25.zip
   creating: /usr/local/data-integration/
   creating: /usr/local/data-integration/system/
   creating: /usr/local/data-integration/system/karaf/
   creating: /usr/local/data-integration/system/karaf/system/
   creating: /usr/local/data-integration/system/karaf/system/org/
   creating: /usr/local/data-integration/system/karaf/system/org/apache/
   creating: /usr/local/data-integration/system/karaf/system/org/apache/karaf/
   creating: /usr/local/data-integration/system/karaf/system/org/apache/karaf/bu
ndle/
   creating: /usr/local/data-integration/system/karaf/system/org/apache/karaf/bu
ndle/org.apache.karaf.bundle.command/
   creating: /usr/local/data-integration/system/karaf/system/org/apache/karaf/co
nfig/
   creating: /usr/local/data-integration/system/karaf/system/org/apache/karaf/co
nfig/org.apache.karaf.config.command/
   creating: /usr/local/data-integration/system/karaf/system/org/apache/karaf/di
agnostic/
   creating: /usr/local/data-integration/system/karaf/system/org/apache/karaf/di
agnostic/org.apache.karaf.diagnostic.command/
   creating: /usr/local/data-integration/system/karaf/system/org/apache/karaf/fe
```

图 1-21　解压 Kettle 文件到相应目录

解压完成后进入/usr/local，将文件“data-integration”重命名为“kettle”，以方便后续使用，如图 1-22 所示。

（2）配置环境变量

打开命令行窗口，输入“sudo vim ～/.bashrc”，配置环境变量，如图 1-23 所示。

```
hadoop@ubuntu: /usr/local
hadoop@ubuntu:~/Downloads$
hadoop@ubuntu:~/Downloads$ cd /usr/local
hadoop@ubuntu:/usr/local$ ls
bin               myeclipse-2017-ci-10-offline-installer-linux.zip.qbl
data-integration  pdi-ce-7.0.0.0-25.zip
etc               pig
flume             pycharm
games             sbin
hadoop            share
include           src
java              test
lib               testflume
                  tomcat
mongodb
hadoop@ubuntu:/usr/local$ sudo mv data-integration/ kettle
hadoop@ubuntu:/usr/local$ ls
bin      kettle                                                    pycharm
etc      lib                                                       sbin
flume                                                              share
games    mongodb                                                   src
hadoop   myeclipse-2017-ci-10-offline-installer-linux.zip.qbl      test
include  pdi-ce-7.0.0.0-25.zip                                     testflume
java     pig                                                       tomcat
hadoop@ubuntu:/usr/local$
```

图 1-22　文件的重命名

```
# 配置Kettle环境
export KETTLE_HOME=/usr/local/kettle

# 配置Java环境
export JAVA_HOME=/usr/local/java/jdk1.8.0_171
export JRE_HOME=${JAVA_HOME}/jre
export CLASSPATH=.:${JAVA_HOME}/lib:${JRE_HOME}/lib
export PATH=${JAVA_HOME}/bin:$PATH
```

图 1-23　配置 Kettle 的环境变量

环境变量配置完成并保存后，执行“Source ～/.bashrc”命令，使配置的环境变量生效。

（3）验证 Kettle 是否安装成功

打开命令行窗口，切换到/usr/local/kettle 路径，执行“./spoon.sh”命令，若显示信息如图 1-24 和图 1-25 所示，则表示安装成功。

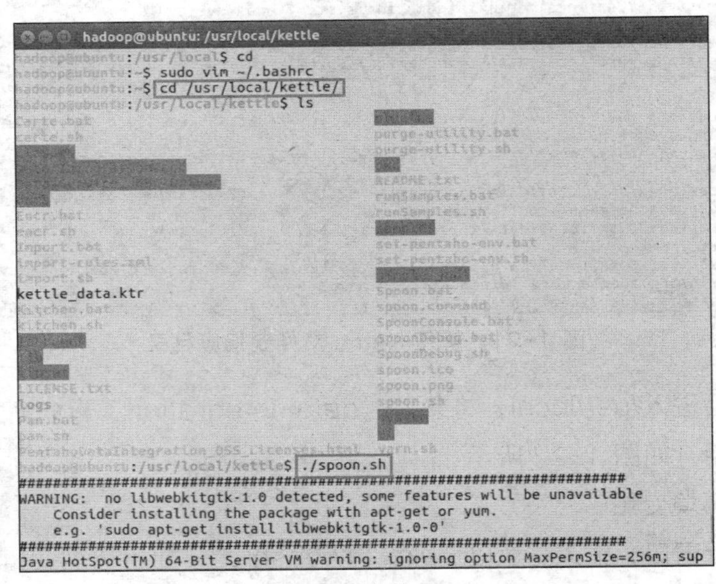

图 1-24　启动 Kettle

Spoon - [no name]
Connect
View
Design
Explorer
Transformations
Jobs

图 1-25　启动成功

项目 2

网络爬虫实践

学习目标

【知识目标】

① 识记网络爬虫的结构。
② 熟悉网络爬虫的基础知识。

【技能目标】

① 熟悉爬虫基本库 urllib 和 Selenium 的使用。
② 熟悉 BeautifulSoup 和 PyQuery 解析库的使用。
③ 学会编写网络爬虫采集北京公交线路相关信息的方法。

项目描述

互联网已经成为信息的主要载体，因此“互联网+”时代下的市场数据分析工作，越来越看重从各类网页中获取、梳理数据信息。那么，如何在互联网中高效搜集到全面而有效的信息呢？本项目就来学习网络爬虫（Web Crawler）技术。

网络爬虫是一种按照一定的规则，自动地抓取万维网信息的程序或脚本，它们被广泛用于互联网搜索引擎或其他类似网站，可以自动采集所有其能够访问到的页面内容，以获取或更新这些网站的内容和检索方式。从功能上来讲，爬虫一般分为数据采集、处理和存储 3 部分。

任务 1　使用 urllib 爬取北京公交线路信息

任务描述

（1）学习网络爬虫相关技术，熟悉爬虫基本库 urllib 的使用。
（2）熟悉网络爬虫相关基础知识。

（3）使用 urllib 基本库获取北京公交线路信息的 HTML 源代码。

（4）使用 BeautifulSoup 解析库完成北京公交线路相关信息的获取。

任务目标

（1）知道 urllib 基本库和 BeautifulSoup 解析库的使用方法。

（2）学会使用 urllib 基本库和 BeautifulSoup 解析库进行北京公交线路相关信息的爬取。

知识准备

1. HTTP 的理解

（1）URL 和 URI

统一资源定位符（Uniform Resource Locator，URL）和统一资源标识符（Uniform Resource Identifier，URI）用于精确地说明某资源的位置以及如何去访问它。URI 有两种表现形式——URL 和统一资源名称（Uniform Resource Name，URN）。URL 描述了一台特定服务器上某资源的特定位置；URN 仅命名资源而不指定如何定位资源，在目前的互联网中使用得很少。因此，几乎所有的 URI 都是 URL。

（2）HTTP 和 HTTPS

在访问百度网页的 URL（HTTPS://www.baidu.com/）时，URL 前部一般可以看到 HTTP 或 HTTPS，这就是访问资源需要的协议类型，有时还可以使用 FTP、SFTP、SMB 等协议类型。爬虫常用的协议类型就是 HTTP 和 HTTPS。

超文本传输协议（Hyper Text Transfer Protocol，HTTP）是一个客户端和服务器端请求和应答的标准，是互联网中应用最为广泛的一种网络协议，所有的 WWW 文件都要遵守这个协议，目前广泛使用的是 HTTP 1.1。

HTTPS 是以安全为目标的 HTTP 通道，可以理解为 HTTP 的安全版，即 HTTP 中加入 SSL 层，其传输的内容都是经过 SSL 加密的，它的主要作用如下：建立一个信息安全通道，来保证数据的传输安全；确认网站的真实性，凡是使用了 HTTPS 的网站，用户都可以通过单击浏览器地址栏中的锁头标志来查看网站认证之后的真实信息，也可以通过证书颁发机构颁发的安全签章来查询。

（3）HTTP 请求流程

一次 HTTP 操作称为一个事务，其工作过程可分为以下 4 步。

首先，在浏览器地址栏中输入一个地址（或单击一个超链接），HTTP 的工作就开始了。

其次，建立连接后，客户端发送一个 HTTP 请求到服务器，请求消息由请求行、请求头部、空行和请求数据 4 部分组成。图 2-1 给出了请求报文的一般格式。

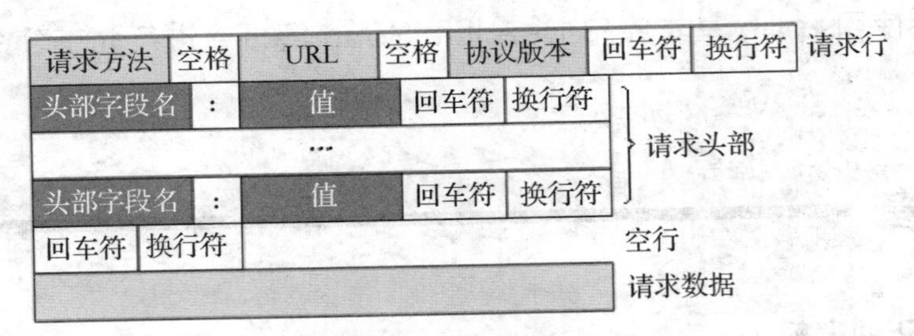

图 2-1 请求报文的一般格式

再次，服务器接收到请求后，给出响应，HTTP 响应也由 4 部分组成，分别是状态行、消息报头、空行和响应正文。

最后，客户端浏览器接收响应之后，在用户的浏览器上渲染显示，客户端和服务器端断开连接。

2. 网页基础知识

（1）网页的组成

网页的组成可分为三大部分：超文本标记语言（Hyper Text Markup Language，HTML）、层叠样式表（Cascading Style Sheets，CSS）和 Java 脚本（JavaScript，JS）。HTML 负责语义，CSS 负责样式，JavaScript 负责交互和行为。

HTML：用来描述网页的一种语言。可以通过 Chrome 浏览器打开一个网址，右键单击检查或按 F12 键，打开“开发者工具”，选择“Elements”选项卡，即可看到网页的源代码。

CSS：“层叠”是指当在 HTML 中引用了数个样式文件，并且样式发生冲突时，浏览器能依据层叠顺序处理；“样式”指网页中文字大小、颜色、元素间距、排列等格式。

JavaScript：一种脚本语言。HTML 和 CSS 配合使用，提供给用户的只是一种静态的信息，缺少交互性。有时在网页中可能会看到一些交互和动画效果，如下载进度条、提示框、轮播图等，这通常需要使用 JavaScript 来实现。

网页的基本结构如下。

```
<!DOCTYPE html>
<html>
<head>
        <meta charset="utf-8">
<title>HelloWorld</title>
</head>
<body>
        <p>Hello World</p>
        <div class="content">
            主体内容
</div>
</body>
</html>
```

一般而言，网页的首行标识 HTML 版本，一对 html 标签包裹 head 和 body 标签，head 标签通常存放一些配置和资源引用，body 标签则存放网页的主体内容。

（2）节点树及节点间的关系

在网页中，组织页面的对象被渲染成一个树形结构，用来表示文档中对象的标准模型，称为文档对象模型（Document Object Model，DOM）。DOM 实际上是以面向对象方式描述的文档模型。DOM 定义了表示和修改文档所需的对象、对象的行为和属性，以及对象之间的关系。可以把 DOM 看作页面上数据和结构的一个树形表示，只是页面可能并不是以这种树的方式具体实现的。节点树示意图如图 2-2 所示。

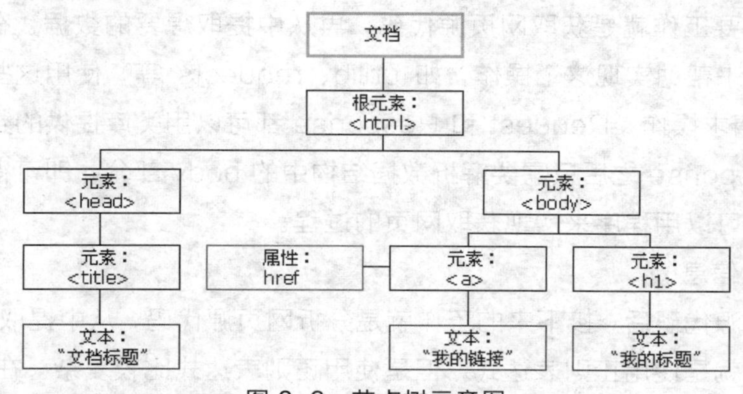

图 2-2　节点树示意图

DOM 规定：整个文档是一个文档节点，每个 HTML 标签是一个元素节点，包含在 HTML 元素中的文本是文本节点，每一个 HTML 属性是一个属性节点，注释属于注释节点。节点树中的节点彼此拥有层级关系。父（Parent）、子（Child）和兄弟（Sibling）等术语用于描述这些关系。父节点拥有子节点。同级的子节点被称为兄弟（同胞或姐妹）。

① 在节点树中，顶端节点被称为根。

② 每个节点都有父节点，除了根（即根节点，没有父节点）。

③ 一个节点可拥有任意数量的子节点。

④ 兄弟是拥有相同父节点的节点。

图 2-3 展示了节点树的一部分及节点之间的关系。

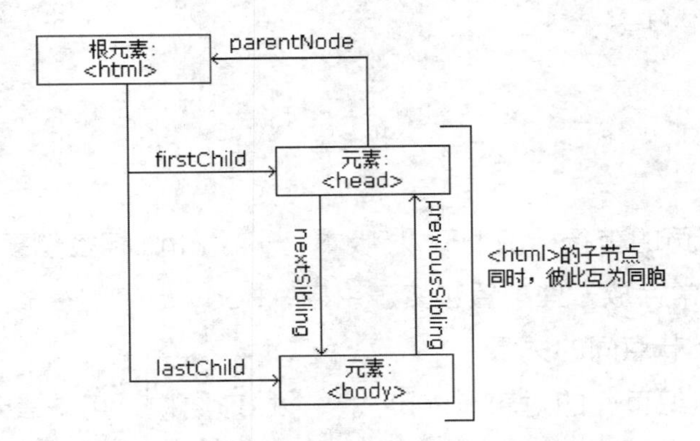

图 2-3　节点树的一部分及节点之间的关系

3. 爬虫的基本原理

网络爬虫（又被称为网页蜘蛛）本质上就是获取网页并提取和保存信息的自动化程序。

（1）获取网页

爬虫的首要工作就是获取网页源代码，再从中提取想要的数据。在 Python 中提供了许多库来帮助实现这个操作，如 urllib、requests 等，使用这些库可以帮助实现 HTTP 请求操作，Request 和 Response 都可以用类库提供的数据结构来表示，得到 Response 之后只需要解析数据结构中的 body 部分，即可得到网页的源代码，这样便可以用程序来实现获取网页的过程。

（2）提取信息

获取网页源代码后，接下来的工作就是分析网页源代码，从中提取想要的数据。最通用的方法就是使用正则表达式，但是使用正则表达式比较复杂。在 Python 中，使用 BeautifulSoup、PyQuery、LXML 等库，可以帮助用户高效地从源代码中提取网页信息。

（3）保存数据

提取信息之后，可以将数据保存到本地，以便后续使用。保存方式有很多种，如 TXT、JSON，也可以保存到数据库中，如 MySQL、MongoDB 等。

注意

现在越来越多的网页使用 JS 来构建，使用 urllib、requests，只能得到静态的 HTML 源代码，它不能加载 JS 文件，对于这类情况，可以分析后台 Ajax 接口，也可以借助 Selenium、Splash 等库来实现模拟 JavaScript 渲染，以便爬取 JavaScript 渲染的网页内容。

4. 基本库的使用

（1）urllib 库的使用

urllib 库是 Python 中的一个功能强大、用于操作 URL，并在制作爬虫的时候经常要用到的库。同样的库还有 requests、httplib2。在 Python 2 中，分别有 urllib 和 urllib 2，但在 Python 3 中，其统一合并到 urllib 中。相对来说，Python 3 对中文的支持比 Python 2 友好，所以下面通过 Python 3 来介绍 urllib 库的一些常见用法。

① 发送请求

```
import urllib.request
r=urllib.request.urlopen(<a rel="external nofollow" href="HTTP://www.python.org/">
HTTP://www.python.org/</a>)
```

其先导入 urllib.request 模块，使用 urlopen()对参数中的 URL 发送请求，返回一个 HTTP.client.HTTPResponse 对象。

在 urlopen()中，使用 timeout 字段，可设定相应的秒数时间之后停止等待响应。除此之外，还可使用 r.info()、r.getcode()、r.geturl()获取相应的当前环境信息、状态码、当前网页 URL。

② 读取响应内容

```
import urllib.request
url = "HTTP://www.python.org/"
with urllib.request.urlopen(url) as r:
r.read()
```

使用 r.read()读取响应内容到内存中，该内容为网页的源代码（可用相应浏览器的“查看网页源代码”功能查看），并可对返回的字符串进行相应解码，即 decode()。

③ 传递 URL 参数

```
import urllib.request
import urllib.parse
params = urllib.parse.urlencode({'q': 'urllib', 'check_keywords': 'yes', 'area': 'default'})
url = "HTTPS://docs.python.org/3/search.html?{}".format(params)
r = urllib.request.urlopen(url)
```

以字符串字典的形式，通过 urlencode()编码，为 URL 的查询字符串传递数据，编码后的 params 为字符串，字典每项键值对以'&'连接，如'q=urllib&check_keywords=yes&area=default'，构建后的 URL 为 HTTPS://docs.python.org/3/search.html?q=urllib&check_keywords=yes&area=default。

当然，urlopen()支持直接构建的 URL，简单的 GET 请求可以不通过 urlencode()编码，手动构建后直接请求即可。

④ 传递中文参数

```
import urllib.request
searchword = urllib.request.quote(input("请输入要查询的关键字："))
url = "HTTPS://cn.bing.com/images/async?q={}&first=0&mmasync=1".
format(searchword)
r = urllib.request.urlopen(url)
```

该 URL 利用“bing”图片接口，查询关键字“q”的图片。如果直接将中文传入 URL 进行请求，会导致编码错误。因此需要使用 quote()对该中文关键字进行 URL 编码，相应的，可以使用 unquote()进行解码。

⑤ 定制请求头

```
import urllib.request
url = 'HTTPS://docs.python.org/3/library/urllib.request.html'
headers = {
    'User-Agent' : 'Mozilla/5.0 (Windows NT 10.0; Win64; x64) AppleWebKit/537.36
(KHTML, like Gecko) Chrome/61.0.3163.100 Safari/537.36',
    'Referer': 'HTTPS://docs.python.org/3/library/urllib.html'
}
```

```
req = urllib.request.Request(url, headers=headers)
r = urllib.request.urlopen(req)
```

有时爬取一些网页时，会出现 403 错误（Forbidden），即禁止访问。这是因为网站服务器会对访问者的 headers 属性进行身份验证。例如，通过 urllib 库发送的请求，默认以“Python-urllib/X.Y”作为 User-Agent，其中，X 为 Python 的主版本号，Y 为副版本号。所以，需要通过 urllib.request.Request()构建 Request 对象，传入字典形式的 headers 属性，模拟浏览器。

相应的 headers 信息，可通过浏览器的开发者调试工具，即“检查”功能的“Network”标签查看相应的网页得到，或使用抓包分析软件 Fiddler、Wireshark 得到。

除上述方法外，还可以使用 urllib.request.build_opener()或 req.add_header()定制请求头。

在 Python 2 中，urllib 模块和 urllib 2 模块通常一起使用，因为 urllib.urlencode()可以对 URL 参数进行编码，而 urllib2.Request()可以构建 Request 对象，定制请求头，并统一使用 urllib2.urlopen()发送请求。

⑥ 传递 POST 请求

```
import urllib.request
import urllib.parse
url = 'HTTPS://passport.cnblogs.com/user/signin?'
post = {
    'username': 'xxx',
    'password': 'xxxx'
}
postdata = urllib.parse.urlencode(post).encode('utf-8')
req = urllib.request.Request(url, postdata)
r = urllib.request.urlopen(req)
```

在进行注册、登录等操作时，会通过 POST 表单传递信息。此时，首先，需要分析页面结构，构建表单数据 post，使用 urlencode()进行编码处理，返回字符串；其次，需要指定“UTF-8”的编码格式，这是因为 postdata 只能是 bytes 或者 file object；最后，通过 Request()传递 postdata，使用 urlopen()发送请求。

⑦ 下载远程数据到本地

```
import urllib.request
url = "HTTPS://www.python.org/static/img/python-logo.png"
urllib.request.urlretrieve(url, "python-logo.png")
```

爬取图片、视频等远程数据时，可使用urlretrieve()将其下载到本地。其第一个参数为要下载的 URL，第二个参数为下载后的存放路径。例如，下载 Python 官网 Logo 到当前目录中，返回元组（filename, headers）。

⑧ 设置代理 IP 地址

```
import urllib.request
url = "HTTPS://www.cnblogs.com/"
proxy_ip = "180.106.16.132:8118"
proxy = urllib.request.ProxyHandler({'HTTP': proxy_ip})
opener = urllib.request.build_opener(proxy, urllib.request.HTTPHandler)
urllib.request.install_opener(opener)
r = urllib.request.urlopen(url)
```

有时，频繁地爬取一个网页会被网站服务器屏蔽 IP 地址。此时，可通过上述方法设置代理 IP 地址。首先，通过网上代理 IP 地址的网站找一个可以用的 IP 地址，构建 ProxyHandler() 对象，将 HTTP 和代理 IP 地址以字典形式作为参数传入，设置代理服务器的信息。再构建 opener 对象，将 proxy 和 HTTPHandler 类传入。通过 install_opener() 将 opener 设置成全局对象，当使用 urlopen()发送请求时，会使用之前设置的信息发送相应的请求。

⑨ 异常处理

```
import urllib.request
import urllib.error
url = "HTTP://www.w3school.com.cn/"
try:
    r = urllib.request.urlopen(url)
except urllib.error.URLError as e:
    if hasattr(e, 'code'):
        print(e.code)
    if hasattr(e, 'reason'):
        print(e.reason)
```

可以使用 URLError 类处理 URL 的相关异常。导入 urllib.error，捕获 URLError 异常后，因为只有发生 HTTPError 异常（URLError 的子类）时，才会有异常状态码 e.code，所以需要判断异常是否有属性 code。

⑩ Cookie 的使用

```
import urllib.request
import HTTP.cookiejar
url = "HTTP://www.w3school.com.cn/"
cjar = HTTP.cookiejar.CookieJar()
opener = urllib.request.build_opener(urllib.request.HTTPCookieProcessor(cjar))
urllib.request.install_opener(opener)
r = urllib.request.urlopen(url)
```

通过无状态协议 HTTP 访问网页时，Cookie 会维持会话间的状态。例如，有些网站需要登录操作，第一次操作时可通过提交 POST 表单来登录，当爬取该网站中的其他站点时，可以使用 Cookie 来保持登录状态，而不用每次都通过提交表单来登录。

首先，构建 CookieJar()对象 cjar；其次，使用 HTTPCookieProcessor() 处理器处理 cjar；再次，通过 build_opener() 构建 opener 对象，并将其设置成全局的；最后，通过 urlopen() 发送请求。

（2）BeautifulSoup

BeautifulSoup 提供一些简单的、Python 式的函数来处理导航、搜索、修改分析树等功能。它是一个工具箱，通过解析文档为用户提供需要抓取的数据。

BeautifulSoup 自动将输入文档转换为 Unicode 编码，将输出文档转换为 UTF-8 编码。一般而言，在处理时不需要考虑编码方式，除非文档没有指定一个编码方式。此时，BeautifulSoup 无法自动识别编码方式，需要说明原始的编码方式。

① 创建 BeautifulSoup 对象

必须先导入 bs4 库，代码如下。

```
from bs4 import BeautifulSoup
```

创建一个字符串，代码如下。

```
html = """
<html><head><title>hello hadoop</title></head>
<body>
<p class="title" name="hadoop"><b>hello hadoop</b></p>
```

```
<p class="story">there were three Elephant
<a href="HTTP://example.com/test1" class="elephant" id="elephant1"><!-- test1 -->
</a>,
<a href="HTTP://example.com/test2" class="elephant" id="elephant2">test2</a> and
<a href="HTTP://example.com/test3" class="elephant" id="elephant3">test3</a>;
and they lived in the zoo.</p>
<p class="story">...</p>
"""
```

创建 BeautifulSoup 对象，代码如下。

```
soup = BeautifulSoup(html)
```

另外，还可以用本地 HTML 文件来创建对象，代码如下。

```
1soup = BeautifulSoup(open('index.html'),lxml)
```

下面来输出 soup 对象的内容，使用格式化输出方式，代码如下。

```
print soup.prettify()
```

输出结果如下。

```
<html>
<head>
    <title>
     hello hadoop
    </title>
```

② 四大对象种类

BeautifulSoup 将复杂 HTML 文档转换成一个复杂的树形结构，每个节点都是 Python 对象，所有对象都可以归纳为 4 种：Tag、NavigableString、BeautifulSoup、Comment。

a. Tag

它就是 HTML 中的一个个标签，代码如下。

```
<title>hello hadoop</title>

<a class="elephant" href="HTTP://example.com/test1" id="elephant1">test1</a>
```

这里的 title 和 a 标签加上其中包括的内容就是 Tag。下面尝试用 BeautifulSoup 方便地获取 Tag。以下每一段代码的注释部分即为运行结果。

```
print soup.title
```

```
#<title>hello hadoop</title>

print soup.head
#<head><title>hello hadoop</title></head>

print soup.a
#<a class="elephant" href="HTTP://example.com/test1" id="elephant1"><!-- test1 -->
</a>

print soup.p
#<p class="title" name="hadoop"><b>hello hadoop</b></p>
```

可以利用 soup 加标签名轻松地获取这些标签的内容，这里需要注意，其查找的是所有内容中的第一个符合要求的标签（查询所有标签的方法稍后会进行介绍）。

下面来验证一下这些对象的类型，代码如下。

```
print type(soup.a)
#<class 'bs4.element.Tag'>
```

对于 Tag，它有两个重要的属性，分别是 name 和 attrs。

```
print soup.name
print soup.head.name
#[document]
#head
```

soup 对象本身比较特殊，它的 name 即为[document]，对于其他内部标签，其输出的值便为标签本身的名称。这里，输出了 p 标签的所有属性，得到的类型是一个字典。

```
print soup.p.attrs
#{'class': ['title'], 'name': 'hadoop'}
```

如果想要单独获取某个属性，则可以按以下方法来做，如获取对象的 class。

```
print soup.p['class']
#['title']
```

还可以利用 get 方法，传入属性的名称来获取对象的 class。

```
print soup.p.get('class')
#['title']
```

此外，可以对这些属性和内容等进行修改，代码如下。

```
soup.p['class']="newClass"
print soup.p
#<p class="newClass" name="hadoop"><b>hello hadoop</b></p>
```

b. NavigableString

不仅可以得到标签的内容，还可以通过“.string”获取标签内部的文字，代码如下。

```
print soup.p.string
#hello hadoop
```

这样可以轻松地获取标签中的内容，下面来检查其类型。

```
print type(soup.p.string)
#<class 'bs4.element.NavigableString'>
```

c. BeautifulSoup

BeautifulSoup对象表示的是一个文档的全部内容，大部分时候，可以将其当作一个特殊的 Tag 对象，可以分别获取它的类型、名称和属性。

```
print type(soup.name)
#<type 'str'>
print soup.name
# [document]
print soup.attrs
#{} 空字典
```

d. Comment

Comment 对象是一个特殊类型的 NavigableString 对象，其输出的内容仍然不包括注释符号，但是如果不妥善地对其进行处理，则可能会给文本处理带来意想不到的麻烦。下面来查找一个带注释的标签。

```
print soup.a
print soup.a.string
print type(soup.a.string)
```

其运行结果如下。

```
<a class="elephant" href="HTTP://example.com/test1" id="elephant1"><!-- test1 --></a>
test1
<class 'bs4.element.Comment'>
```

a 标签中的内容实际上是注释，但是如果利用.string 来输出其内容，将会发现它已经把注释符号去掉了，这样做可能会带来严重的问题。

另外，输出其类型时，会发现它是 Comment 类型，所以，在使用前最好进行判断，代码如下。

```
if type(soup.a.string)==bs4.element.Comment:
        print soup.a.string
```

在上面的代码中，先判断了它的类型是否为 Comment，再进行其他操作，如输出。

③ 遍历文档树

a. 遍历直接子节点

Tag 的.contents 属性可以使 Tag 的子节点以列表的方式输出。

```
print soup.head.contents
#[<title>hello hadoop</title>]
```

输出方式为列表，可以用列表索引来获取它的某一个元素。

```
print soup.head.contents[0]
#<title>hello hadoop</title>
```

.children 返回的不是一个 list，而是一个 list 生成器对象，可以通过遍历获取所有子节点。

```
for child in soup.body.children:
        print child


#<p class="title" name="hadoop"><b>hello hadoop</b></p>
#<p class="story">there were three Elephants
#<a class="elephant" href="HTTP://example.com/test1" id="elephant1"><!-- test1 -->
</a>,
#<a class="elephant" href="HTTP://example.com/test2" id="elephant2">test2</a> and
<a class="elephant" href="HTTP://example.com/test3" id="elephant3">test3</a>;
and they lived in the zoo.</p>
<p class="story">...</p>
```

b. 遍历所有子孙节点

.contents 和.children 属性仅包含 Tag 的直接子节点，.descendants 属性可以对所有 Tag 的子孙节点进行递归循环，其和.children 类似，需要遍历获取其中的内容。

```
for child in soup.descendants:
```

```
        print child
```

其运行结果如下。可以发现，所有的节点都被打印出来了，先是最外层的 html 标签，再从 head 中将标签一个个剥离出来，以此类推。

```
#<html><head><title>hello hadoop</title></head>
#<body>
#<p class="title" name="hadoop"><b>hello hadoop</b></p>
#<p class="story">there were three Elephants
<a class="elephant" href="HTTP://example.com/test1" id="elephant1"><!-- test1 -->
</a>,
```

c. 遍历节点内容

如果 Tag 只有一个 NavigableString 类型的子节点，那么这个 Tag 可以使用.string 得到子节点。如果一个 Tag 仅有一个子节点，那么这个 Tag 也可以使用.string 方法，其输出结果与当前唯一子节点的.string 的输出结果相同。

简单来讲，如果一个标签中没有标签了，那么.string 会返回标签中的内容。如果标签中只有唯一的一个标签，那么.string 会返回最里面的内容。

```
print soup.head.string
#hello hadoop
print soup.title.string
#hello hadoop
```

如果 Tag 包含了多个子节点，则 Tag 无法确定.string 方法应该调用哪个子节点的内容，.string 的输出结果是 None。

```
print soup.html.string
# None
```

d. 遍历多个内容

.string 可以通过遍历来获取多个内容，代码如下。

```
for string in soup.strings:
        print(repr(string))
        # u"hello hadoop"
        # u'\n\n'
        # u"hello hadoop"
        # u'\n\n'
        # u'there were three Elephants\n'
```

```
        # u'test1'
        # u',\n'
        # u'test2'
        # u' and\n'
        # u'test3'
        # u';\nand they lived in the zoo.'
        # u'\n\n'
        # u'...'
        # u'\n'
```

其输出的字符串中可能包含了很多空格或空行，使用.stripped_strings 可以去除多余空白内容，代码如下。

```
for string in soup.stripped_strings:
        print(repr(string))
        # u"hello hadoop"
        # u"hello hadoop"
        # u'there were three Elephants'
        # u'test1'
        # u','
        # u'test2'
        # u'and'
        # u'test3'
        # u';\nand they lived in the zoo.'
        # u'...'
```

e. 遍历父节点

使用元素的 .parent 属性可以获取父节点，代码如下。

```
p = soup.p
print p.parent.name
#body

content = soup.head.title.string
print content.parent.name
#title
```

f. 遍历全部父节点

通过元素的.parents 属性可以递归得到元素的所有父节点，代码如下。

```
content = soup.head.title.string
for parent in  content.parents:
        print parent.name


#title
#head
#html
#[document]
```

g. 遍历兄弟节点

兄弟节点可以理解为和本节点处在同一级的节点，.next_sibling 属性可获取该节点的下一个兄弟节点，.previous_sibling 则与之相反，如果节点不存在，则返回 None。

> **注意** **实际文档中的 Tag 的.next_sibling 和.previous_sibling 属性通常是字符串或空白的，因为空白或者换行也可以被视作一个节点，所以得到的结果可能为空白或者换行。**

```
print soup.p.next_sibling
#该处有一个空白行


#因为前一个兄弟节点为空白行，所以返回 None
print soup.p.previous_sibling
print soup.p.next_sibling.next_sibling
#<p class="story">there were three elephants
#<a class="elephant" href="HTTP://example.com/test1" id="elephant1"><!-- test1 --></a>,
#<a class="elephant" href="HTTP://example.com/test2" id="elephant2">test2</a> and
#<a class="elephant" href="HTTP://example.com/test3" id="elephant3">test3</a>;
#and they lived in the zoo.</p>
#下一个节点的下一个兄弟节点是可以看到的节点
```

h. 遍历全部兄弟节点

通过 .next_siblings 和 .previous_siblings 属性可以对当前节点的兄弟节点进

行迭代输出。

```
for sibling in soup.a.next_siblings:
        print(repr(sibling))
        # u',\n'
        # <a class="elephant" href="HTTP://example.com/test2" id="elephant2">test2</a>
        # u' and\n'
        # <a class="elephant" href="HTTP://example.com/test3" id="elephant3">test3</a>
        # u'; and they lived in the zoo.'
        # None
```

i. 遍历前后节点

与.next_sibling 和.previous_sibling 不同，.next_element 并不是针对兄弟节点，而是针对所有节点，不分层次，如 head 节点。

```
<head><title>hello hadoop</title></head>
```

那么 head 节点的下一个节点便是 title，它是不分层次关系的。

```
print soup.head.next_element
#<title>hello hadoop</title>
```

j. 遍历所有前后节点

通过.next_elements 和.previous_elements 的迭代器可以向前或向后访问文档的解析内容，就好像文档正在被解析一样。

```
for element in last_a_tag.next_elements:
        print(repr(element))
# u'test3'
# u';\nand they lived in the zoo.'
# u'\n\n'
# <p class="story">...</p>
# u'...'
# u'\n'
# None
```

④ 搜索文档树

a. find_all(name,attrs,recursive,text,limit,**kwargs)

find_all()方法用于搜索当前 Tag 的所有 Tag 子节点，并判断是否符合过滤器的条件。

（a）name 参数

name 参数可以查找所有名称为 name 的 tag，字符串对象会被自动忽略。

ⓐ 传字符串：最简单的过滤器是字符串。在搜索方法中传入一个字符串参数，BeautifulSoup 会查找与字符串完整匹配的内容。下面的例子用于查找文档中所有的 b 标签。

```
soup.find_all('b')
# [<b>hello hadoop</b>]

print soup.find_all('a')
#[<a class="elephant" href="HTTP://example.com/test1" id="elephant1"><!-- test1 -->
</a>, <a class="elephant" href="HTTP://example.com/test2" id="elephant2">test2</a>,
<a class="elephant" href="HTTP://example.com/test3" id="elephant3">test3</a>]
```

ⓑ 传正则表达式：如果传入正则表达式作为参数，则 BeautifulSoup 会通过正则表达式的 match() 来匹配内容。下面例子用于找出所有以 b 开头的标签，这表示 body 和 b 标签都应该被找到。

```
import re
for tag in soup.find_all(re.compile("^b")):
    print(tag.name)
# body
# b
```

ⓒ 传列表：如果传入列表参数，则 BeautifulSoup 会将与列表中任一元素匹配的内容返回。下面的例子用于找到文档中所有 a 标签和 b 标签。

```
soup.find_all(["a", "b"])
# [<b>hello hadoop</b>,
#    <a class="elephant" href="HTTP://example.com/test1" id="elephant1">test1</a>,
#    <a class="elephant" href="HTTP://example.com/test2" id="elephant2">test2</a>,
#    <a class="elephant" href="HTTP://example.com/test3" id="elephant3">test3</a>]
```

ⓓ 传 True：True 可以匹配任何值。下面的例子用于查找到所有的 Tag，但是不会返回字符串节点。

```
for tag in soup.find_all(True):
    print(tag.name)
# html
```

```
# head
# title
# body
# p
# b
# p
# a
# a
```

ⓔ 传方法：如果没有合适的过滤器，还可以定义一个方法，此方法只接收一个元素参数（HTML 文档中的一个元素节点，不能是文本节点），如果这个方法返回 True，则表示当前元素匹配并且被找到，否则返回 False。

下面的例子用于校验当前元素，如果包含 class 属性却不包含 id 属性，那么将返回 True。

```
def has_class_but_no_id(tag):
        return tag.has_attr('class') and not tag.has_attr('id')
```

将这个方法作为参数传入 find_all() 方法，将得到所有 p 标签。

```
soup.find_all(has_class_but_no_id)
# [<p class="title"><b>hello hadoop</b></p>,
#   <p class="story">there were...</p>,
#   <p class="story">...</p>]
```

（b）attrs 参数

有些 Tag 属性在搜索中无法使用，如 HTML 5 中的 data-*属性。

```
data_soup = BeautifulSoup('<div data-foo="value">foo!</div>')
data_soup.find_all(data-foo="value")
# SyntaxError: keyword can't be an expression
```

但是可以通过 find_all() 方法的 attrs 参数定义一个字典参数来搜索包含特殊属性的 Tag，表达式可以是字符串、布尔值、正则表达式。

```
data_soup.find_all(attrs={"data-foo": "value"})
# [<div data-foo="value">foo!</div>]
```

（c）recursive 参数

调用 Tag 的 find_all() 方法时，BeautifulSoup 会检索当前 Tag 的所有子孙节点，如果只想搜索 Tag 的直接子节点，则可以使用参数 recursive=False。

一段简单的文档如下。

```
<html>
<head>
    <title>
     hello hadoop
    </title>
</head>
...
```

是否使用 recursive 参数的搜索结果对比如下。

```
soup.html.find_all("title")
# [<title>hello hadoop</title>]

soup.html.find_all("title", recursive=False)
# []
```

（d）text 参数

通过 text 参数可以搜索文档中的字符串内容。与 name 参数的可选值一样，text 参数可以为字符串、正则表达式、列表和 True。

```
soup.find_all(text="test1")
# []
soup.find_all(text=["test3", "test1", "test2"])
# ['test2', 'test3']
soup.find_all(text=re.compile("hadoop"))
# ["hello hadoop", "hello hadoop"]
```

（e）limit 参数

find_all() 方法返回全部的搜索结果，如果文档树很大，搜索就会很慢。如果不需要全部结果，则可以使用 limit 参数限制返回结果的数量。当搜索到的结果数量达到 limit 的限制时，就停止搜索返回结果。

文档树中有 3 个 Tag 符合搜索条件，但结果只返回了 2 个，因为 limit=2 限制了返回数量。

```
soup.find_all("a", limit=2)
#  [<a class="elephant" href="HTTP://example.com/test1" id="elephant1">test1</a>,
#  <a class="elephant" href="HTTP://example.com/test2" id="elephant2">test2</a>]
```

（f）**kwargs 参数

注意　如果一个指定名称的参数不是搜索内置的参数名，则搜索时会把该参数当作指定名称 Tag 的属性来搜索，如果包含一个名称为 id 的参数，则 BeautifulSoup 会搜索每个 Tag 的 id 属性。

```
soup.find_all(id='elephant2')
#  [<a class="elephant" href="HTTP://example.com/test2" id="elephant2">test2</a>]
```

如果传入 href 参数，则 BeautifulSoup 会搜索每个 Tag 的 href 属性。

```
soup.find_all(href=re.compile("test1"))
#  [<a class="elephant" href="HTTP://example.com/test1" id="elephant1">test1</a>]
```

使用多个指定名称的参数可以同时过滤 Tag 的多个属性。

```
soup.find_all(href=re.compile("test1"), id='elephant1')
#  [<a class="elephant" href="HTTP://example.com/test1" id="elephant1">three</a>]
```

如果想用 class 进行过滤，则由于 class 是 Python 的关键词，因此需要为其加一个下划线来解决冲突问题。

```
soup.find_all("a", class_="elephant")
#  [<a class="elephant" href="HTTP://example.com/test1" id="elephant1">test1</a>,
#    <a class="elephant" href="HTTP://example.com/test2" id="elephant2">test2</a>,
#    <a class="elephant" href="HTTP://example.com/test3" id="elephant3">test3</a>]
```

b．find_parents()与 find_parent()

find_parents() 在当前节点的先辈节点中搜索，返回列表类型，参数同.find_all()参数。而 find_parent() 在当前节点的先辈节点中搜索，则返回一个结果，参数同.find()参数。

c．find_next_siblings()与 find_next_sibling()

find_next_siblings()方法在后续平行节点中搜索，返回列表类型，参数同.find_all()参数。find_next_sibling()方法在后续平行节点中搜索，返回第一个符合条件的结果，参数同.find()参数。

d．find_previous_siblings()与 find_previous_sibling()

find_previous_siblings()方法在前序平行节点中搜索，返回列表类型，参数同.find_all()参数。find_previous_sibling()方法在前序平行节点中搜索，返回第一个符合条件的结果，参数同.find()参数。

e．find_all_next()与 find_next()

find_all_next()方法在当前节点之后的节点中搜索，返回列表类型，参数同.find_all()参数。find_next()方法在当前节点之后的节点中搜索，返回第一个符合条件的节点，参数同.find()参数。

f．find_all_previous()与 find_previous()

find_all_previous()方法在当前节点之前的节点中搜索，返回列表类型，参数同.find_all()参数。find_previous()方法在当前节点之前的节点中搜索，返回第一个符合条件的节点，参数同.find()参数。

⑤ CSS 选择器

编写 CSS 时，标签名不加任何修饰，类名前加点，id 名前加#，在这里也可以利用类似的方法来筛选元素，使用的方法是 soup.select()，返回类型是 list。

a．通过标签名查找

```
print soup.select('title')
#[<title>hello hadoop</title>]

print soup.select('a')
#[<a class="elephant" href="HTTP://example.com/test1" id="elephant1"><!-- test1 -->
</a>, <a class="elephant" href="HTTP://example.com/test2" id="elephant2">test2</a>,
<a class="elephant" href="HTTP://example.com/test3" id="elephant3">test3</a>]

print soup.select('b')
#[<b>hello hadoop</b>]
```

b．通过类名查找

```
print soup.select('.elephant')
#[<a class="elephant" href="HTTP://example.com/test1" id="elephant1"><!-- test1 -->
</a>, <a class="elephant" href="HTTP://example.com/test2" id="elephant2">test2</a>, <a
class="elephant" href="HTTP://example.com/test3" id="elephant3">test3</a>]
```

c．通过 id 名查找

```
print soup.select('#elephant1')
#[<a class="elephant" href="HTTP://example.com/test1" id="elephant1"><!-- test1 -->
```

```
</a>]
```

d. 组合查找

组合查找编写 class 文件时，标签名与类名、id 名进行组合的原理是一样的，例如，查找 p 标签时，id 等于 elephant1 的内容，两者需要用空格分开。

```
print soup.select('p #elephant1')
#[<a class="elephant" href="HTTP://example.com/test1" id="elephant1"><!-- test1 -->
</a>]
```

e. 直接子标签查找

```
print soup.select("head > title")
#[<title>hello hadoop</title>]
```

f. 属性查找

进行查找时，还可以加入属性元素，属性需要用中括号括起来。注意，属性和标签属于同一节点，所以中间不能加空格，否则会无法匹配。

```
print soup.select('a[class="elephant"]')
#[<a class="elephant" href="HTTP://example.com/test1" id="elephant1"><!-- test1 -->
</a>, <a class="elephant" href="HTTP://example.com/test2" id="elephant2">test2</a>, <a
class="elephant" href="HTTP://example.com/test3" id="elephant3">test3</a>]

print soup.select('a[href="HTTP://example.com/test1"]')
#[<a class="elephant" href="HTTP://example.com/test1" id="elephant1"><!-- test1 -->
</a>]
```

同样，属性仍然可以与上述查找方式组合，不在同一节点的用空格隔开，在同一节点的不加空格。

```
print soup.select('p a[href="HTTP://example.com/test1"]')
#[<a class="elephant" href="HTTP://example.com/test1" id="elephant1"><!-- test1 -->
</a>]
```

以上的 select()方法返回的结果都是列表形式，可以遍历形式输出，并用 get_text()方法来获取其内容。

```
soup = BeautifulSoup(html, 'lxml')
print type(soup.select('title'))
print soup.select('title')[0].get_text()
```

```
for title in soup.select('title'):
    print title.get_text()
```

任务实施

1．确定并分析目标网页架构

（1）进入 HTTP://beijing.8684.cn 网页，即进入北京公交查询页面，如图 2-4 所示。

图 2-4　北京公交查询页面

（2）单击“以数字开头”中的各数字，观察 URL 的变化，可发现网页 URL 变化规律为 HTTP://beijing.8684.cn/list+数字。

（3）单击数字“1”，进入 HTTP://beijing.8684.cn/list1 页面，在右键单击页面弹出的快捷菜单中，选择“检查”命令，进入检查页面，如图 2-5 所示。

由图 2-5 可以看出，这个页面并没有包含详细的北京公交信息，需要进一步查找详细信息所在的页面。

（4）使用快捷键“Ctrl+Shift+C”，单击“以汉字/字母开头”下面的“1 路”超链接，以网页源码查询方式显示网页内容，如图 2-6 所示。

图 2-5　检查页面

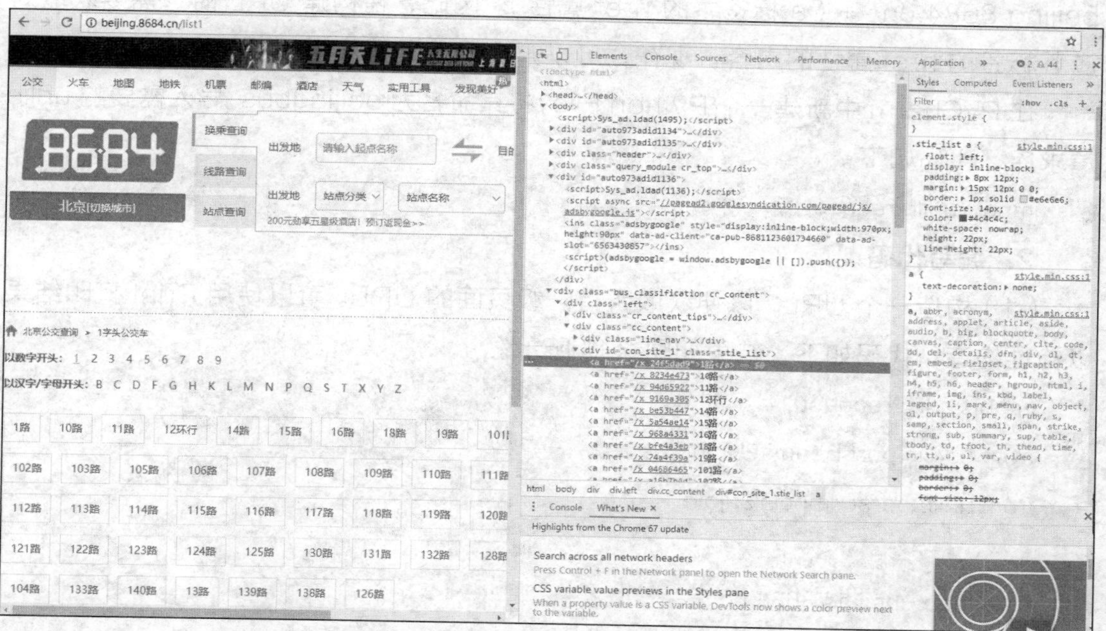

图 2-6　以网页源码查询方式显示网页内容

（5）单击 a 标签中的 href 属性，进入详细信息页面，如图 2-7 所示。

图 2-7　详细信息页面

此时仔细观察 URL 的变化，将会发现详细信息页面的 URL 变化规律为 HTTP://beijing.8684.cn/ +（a 标签中的 href 属性）。因此，在构建 URL 时，需要获取对应 a 标签的 href 属性的内容。现在，可以开始构建爬虫程序了。

在 PyCharm 中新建一个 Python File，将其命名为 beijingbus，导入需要的 urllib 请求库，代码如下。

```
import urllib.request
```

2. 编写代码

（1）构造一个 URL，以便获取所有一级页面的 URL。可以使用 Print 打印结果以验证构造，代码如下，结果如图 2-8 所示。

```
import urllib.request
from bs4 import BeautifulSoup as bs
from urllib.parse import urljoin


url = 'http://beijing.8684.cn'
url_list = url + '/list%d'
for k in range(1,10):
    urls = url_list % k
    print(urls)
```

```
http://beijing.8684.cn/list1
http://beijing.8684.cn/list2
http://beijing.8684.cn/list3
http://beijing.8684.cn/list4
http://beijing.8684.cn/list5
http://beijing.8684.cn/list6
http://beijing.8684.cn/list7
http://beijing.8684.cn/list8
http://beijing.8684.cn/list9

Process finished with exit code 0
```

图 2-8　构造并验证 URL 的结果

（2）创建一个 get_page_url 方法，并在 for 循环中调用代码，结果如图 2-9 所示。

```
import urllib.request
from bs4 import BeautifulSoup as bs
from urllib.parse import urljoin


url = 'http://beijing.8684.cn'
def get_page_url(urls):
    html = urllib.request.urlopen(urls)
    soup = bs(html.read(),'html.parser')
    lu = soup.find('div',id = 'con_site_1')
    hrefs = lu.find_all('a')
    for k in hrefs:
        urls = urljoin(url,k['href'])
        print(urls)
```

```
https://beijing.8684.cn/x_05db550f
https://beijing.8684.cn/x_bf864a1a
https://beijing.8684.cn/x_968a4331
https://beijing.8684.cn/x_b1e20d0f
https://beijing.8684.cn/x_0c88d7d1
https://beijing.8684.cn/x_94d65922
https://beijing.8684.cn/x_1758ec7d
https://beijing.8684.cn/x_7833ac5a
https://beijing.8684.cn/x_832c590b

Process finished with exit code 0
```

图 2-9　获得并拼接详细页面 URL 的结果

上面使用了 urllib 的 request.urlopen 请求方法获得网页的源代码，使用 BeautifulSoup 解析库解析网页源码，查找到需要的二级网页超链接，并将其拼接起来。先用 find()方法查找源码 div 标签中 id 为 con_site_1 的数据，再在找到的数据中用 find_all()方法获得要查找的 URL 片段，并用 urllib.parse 中的 urljoin()模块拼接 URL。可以使用 Print 打印结果以验证构造结果。

（3）在获得了详细页面的 URL 后，即可开始进行详细页面解析。

创建一个在 get_page_url 中调用的 get_page_info 方法，使用 urllib.request.urlopen 请求网页源码，并声明使用 BeautifulSoup 解析源码，代码如下。

```
def get_page_info(urls):
    html = urllib.request.urlopen(urls)
    soup = bs(html.read(),'html.parser')
```

在详细信息页面右键单击所弹出的快捷菜单中，选择“检查”命令，随后使用快捷键“Ctrl+Shift+C”选中要爬取的数据，如图 2-10 所示。

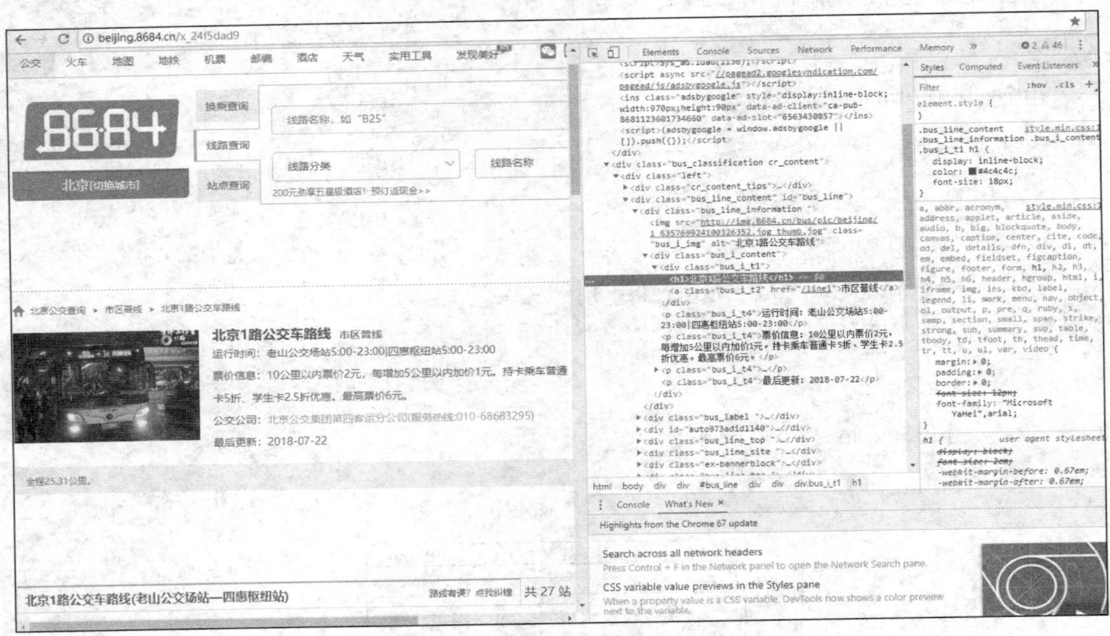

图 2-10　选中要爬取的数据

在要爬取的数据上右键单击，在弹出的快捷菜单中选择“Copy”中的“Copy selector”命令，如图 2-11 所示。

在 get_page_info 方法中，使用 BeautifulSoup 的 select()方法获得需要的公交线路名称、类型、里程等数据，代码如下。

图 2-11 选择“Copy selector”命令

```
def get_page_info(urls):
    html = urllib. request, urlopen(urls)
    soup = bs(html.read()，'html. parser')
    bus_name = soup.select('#bus_line > div. bus_line_information > div > div > hi')[0].
string
    bus_type = soup.select('#bus_line > div. bus_line_information > div > div >
a')[0].string
  try:
    licheng = soup.select('#bus_line > div.bus_label > p')[0].string
  except:
    licheng = None
```

使用 BeautifulSoup 的 find()、find_all()方法爬取更多的相关数据，代码如下。

```
    licheng = None
```

```
bus_time = soup.find('p', class_='bus_i_t4').string
ticket = soup.find_all('p', class_='bus_i_t4')[1].string
gongsi = soup.find_all ('p', class_='bus_i_t4')[2].find_all ('a')[0].string
gengxin = soup.find_all ('p', class_='bus_i_t4')[3].string
try:
    wang_info = soup.find_all('div', class_='bus_line_txt')[0].strong.string
except:
    wang_info = None
try:
  fan_info = soup.find_all('div', class_='bus_line_txt' )[1].strong.string
except:
    fan_info = None
try:
    wang_list_tag = soup.find_all ('div', class_='bus_line_site')[0].find_all ('a')
except:
    wang_list_tag = None
try:
    fan_list_tag = soup.find_all ('div', class_='bus_line_site')[1].find_all('a')
except:
    fan_list_tag=None
```

但是，此时爬取的线路详细信息 wang_list_tag 字段和 fan_list_tag 字段是一些含有标签信息的列表，如图 2-12 所示。

```
[<a href="/z_dbae21df">大屯东</a>, <a href="/z_3075c6d5">大屯南</a>, <a href="/z_cae4a869">炎黄艺术馆</a>, <a href="/z_f251690d">安慧桥北</a>, <a
 href="/z_1f1c8d98">奥体东门</a>, <a href="/z_ad329d3c">五路居</a>, <a href="/z_8f3c46e3">安贞里</a>, <a href="/z_82515a03">安外甘水桥</a>, <a
 href="/z_82ed1070">蒋宅口</a>, <a href="/z_faccdcb0">地坛西门</a>, <a href="/z_bdecfecf">安定门内</a>, <a href="/z_8d0ece65">方家胡同</a>, <a
 href="/z_4f63284c">小经厂</a>, <a href="/z_2d1d26cd">宝钞胡同</a>, <a href="/z_35031adb">鼓楼</a>, <a href="/z_2e7be01b">地安门外</a>, <a
 href="/z_3a657698">地安门内</a>, <a href="/z_62642a29">景山东街</a>, <a href="/z_1d9e1e34">景山东门</a>, <a href="/z_bd3544b2">故宫</a>, <a
 href="/z_7867ec8a">北海</a>, <a href="/z_f9bd5208">西安门</a>, <a href="/z_b5d392e4">西四丁字街</a>]
[<a href="/z_9ae9bc07">安定门</a>, <a href="/z_faccdcb0">地坛西门</a>, <a href="/z_ad348ee7">和平里中街西口</a>, <a href="/z_248b91a4">地坛北门
 </a>, <a href="/z_4debf52c">和平里中街</a>, <a href="/z_e67ce808">和平里北街</a>, <a href="/z_64f6261b">和平西桥南</a>, <a href="/z_0b46af42">和
 平西桥北</a>, <a href="/z_c4f44971">惠新西街南口</a>, <a href="/z_5470a3b0">惠新苑</a>, <a href="/z_a098897a">惠新里</a>, <a
 href="/z_8b439bac">对外经贸大学</a>, <a href="/z_ce6b965c">中国现代文学馆</a>, <a href="/z_be1fd8c5">育慧南路</a>, <a href="/z_9a33dc8a">望和桥
 </a>, <a href="/z_f14ee75f">芍药居</a>]
```

图 2-12　wang_list_tag 字段和 fan_list_tag 字段的列表

要想得到需要的数据，还要遍历列表，此时要使用.string 方法获取标签中的数据，并利用逗号对数据进行拼接处理，即格式化字段，如图 2-13 所示。

```
    fan_list_tag = None
```

```
    wang_buff = ''
    fan_buff = ''
    for wang in wang_list_tag:
        wang_buff += wang.string + ','
    if fan_list_tag:
        for fan in fan_list_tag:
            fan_buff += fan.string + ','
    print(fan_list)
```

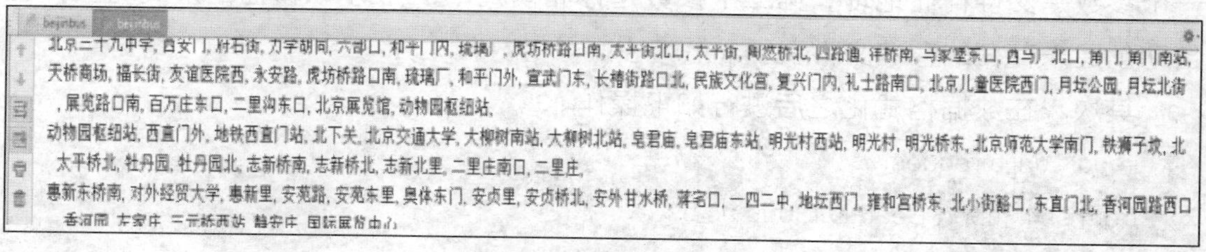

```
北京三十九中学, 西安门, 府右街, 力学胡同, 六部口, 和平门内, 琉璃厂, 虎坊桥路口南, 太平街北口, 太平街, 陶然桥北, 四路通, 洋桥南, 马家堡东口, 西马厂北口, 角门, 角门南站,
天桥商场, 福长街, 友谊医院西, 永安路, 虎坊桥路口南, 琉璃厂, 和平门外, 宣武门东, 长椿街路口北, 民族文化宫, 复兴门内, 礼士路南口, 北京儿童医院西门, 月坛公园, 月坛北街
, 展览路口南, 百万庄东口, 二里沟东口, 北京展览馆, 动物园枢纽站,
动物园枢纽站, 西直门外, 地铁西直门站, 北下关, 北京交通大学, 大柳树南站, 大柳树北站, 皂君庙, 皂君庙东站, 明光村西站, 明光村, 明光桥东, 北京师范大学南门, 铁狮子坟, 北
太平桥北, 牡丹园, 牡丹园北, 志新桥南, 志新桥北, 志新北里, 二里庄南口, 二里庄,
惠新东桥南, 对外经贸大学, 惠新里, 安苑路, 安苑东里, 奥体东门, 安贞里, 安贞桥北, 安外甘水桥, 蒋宅口, 一四二中, 地坛西门, 雍和宫桥东, 北小街豁口, 东直门北, 香河园路西口
香河园, 左家庄, 三元桥西站, 静安庄, 国际展览中心
```

图 2-13　格式化字段

使用两个 list 字段格式化爬取的数据，将第一个 list 中所有数据为空的行置为“None”，即进行空值处理，如图 2-14 所示。

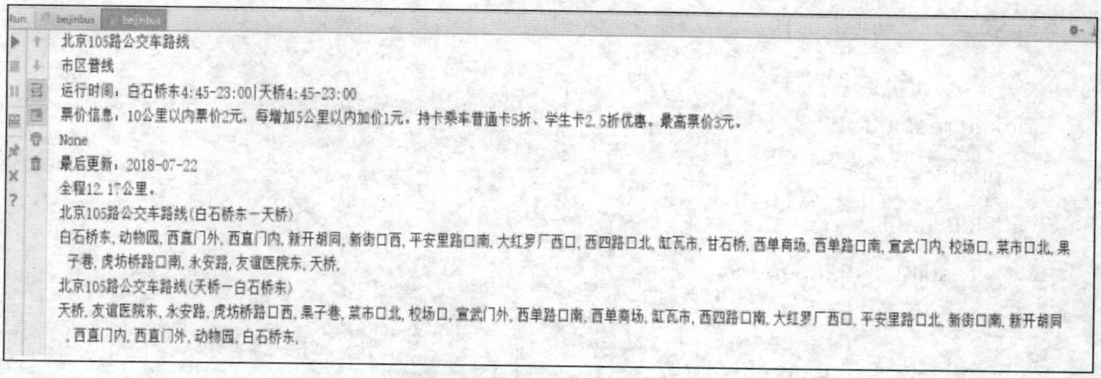

```
北京105路公交车路线
市区普线
运行时间：白石桥东4:45-23:00|天桥4:45-23:00
票价信息：10公里以内票价2元，每增加5公里以内加价1元。持卡乘车普通卡5折、学生卡2.5折优惠。最高票价3元。
None
最后更新：2018-07-22
全程12.17公里。
北京105路公交车路线(白石桥东—天桥)
白石桥东, 动物园, 西直门外, 西直门内, 新开胡同, 新街口西, 平安里路口南, 大红罗厂西口, 西四路口北, 缸瓦市, 甘石桥, 西单商场, 西单路口南, 宣武门内, 校场口, 菜市口北, 果
子巷, 虎坊桥路口南, 永安路, 友谊医院东, 天桥,
北京105路公交车路线(天桥—白石桥东)
天桥, 友谊医院东, 永安路, 虎坊桥路口西, 果子巷, 菜市口北, 校场口, 宣武门外, 西单路口南, 西单商场, 缸瓦市, 西四路口南, 大红罗厂西口, 平安里路口北, 新街口南, 新开胡同
, 西直门内, 西直门外, 动物园, 白石桥东,
```

图 2-14　空值处理

```
    result_list=[bus_name,bus_type,bus_time,trick,gongsi,gengxin,wang_info,fan_info]
    result2_list = []
    for k in result_list:
        print(k)
        if k is not None:
            result2_list.append(k)
```

```
        else:
            result2_list.append(None)
    url_list = url+'/list%d'
    for k in range(1,10):
```

（4）至此，已经获取了想要爬取的数据。下面需要将其保存起来（在本书中，演示了两种存储方式：文件存储方式和 MySQL 关系数据库存储方式）。

① 文件存储方式

使用 csv 库的 writer 方法进行保存。首先，导入 csv 库，并使用 open 方法（第一个参数是文件存储地址，第二个参数是存储模式，w 表示以覆盖的形式写入并且自动创建未存在的存储文件）。

其次，在数据格式化之后进行如下操作。

```
import urllib.request
from bs4 import BeautifulSoup as bs
from urllib.parse import urljoin


url = 'http://beijing.8684.cn'
cs = open('bus_info.csv','w',newline='')
writer = csv.writer(cs)


for k in result_list:
    print(k)
    if k is not None:
        result2_list.append(k)
    else:
        result2_list.append(None)
writer.writerow(result2_list)
```

最后，整理并格式化代码，将 for 循环写在 main 方法中，代码如下。

```
if __name__=='__main__':
    url_list = url+'/list%d'
    for k in range(1,10):
        urls = url_list % k
        get_page_url(urls)
```

② MySQL 关系数据库存储方式

首先，在 Python 中，下载 MySQL 数据库的驱动 pymysql 并导入项目，代码如下。

```
import urllib.request
from bs4 import BeautifulSoup as bs
from urllib.parse import urljoin
import csv
import pymysql

url = 'http://beijing.8684.cn'
cs = open('bus_info.csv', 'w')
writer = csv.writer(cs)
mysql_url=pymysql.connect(host="localhost",user="stu", passwd="stu123", db="studb")
curson = mysql_url.cursor()
curson.execute ("CREATE TABLE stu_businfo (bus_name varchar (1000), bus_type varchar (1000), bus_time varchar (1000), ticket varchar (1000),gongsi varchar (1000), gengxin varchar (1000), licheng varchar (1000), wang_info varchar (1000), wang_buff varchar(1000),fan_info varchar(1000), fan_buff varchar(1000)) DEFAULT CHARSET=UTF8")
mysql_url.commit ()
```

使用 pymysql.connect 声明数据库连接（host 为数据所在地址，user 为数据库用户名，passwd 为数据库用户密码，db 为所要使用的数据库）。使用 cursor 对象建立连接，使用其.execute 方法对数据库进行操作，以创建相应的数据表，并使用.commit()提交数据库操作。

其次，在爬虫中添加一个 save_mysql 方法，代码如下。

```
def save_mysql(data_list): data = tuple(data_list)
curson.execute( 'INSERT INTO stu_businfo (bus_name, bus_type, bus_time,ticket,gongsi, gengxin,licheng,wang_info,wang_buff,fan_info,fan_buff" VALUES (%s, %s, %s, %s, %s, %s, %s, %s, %s, %s, %s) ", data)
mysql_url.commit()
```

save_mysql 方法会将传入的 list 类型数据转换为元组，并插入数据库。

最后，在 get_page_info 方法后添加如下语句，即可实现数据的存储。

```
    result2_list.append(None)
writer.writerow(result2_list)
save_mysql(result2_list)
```

任务 2　使用 Selenium 爬取淘宝网站信息

任务描述

（1）借助学习论坛、网络视频等网络资源和各种图书资源，学习网络爬虫相关技术，熟悉爬虫基本库 Selenium 的使用。

（2）使用 Selenium 基本库获取淘宝网站信息的 HTML 源代码。

（3）使用 PyQuery 解析库实现淘宝网站信息的获取。

任务目标

（1）知道 Selenium 基本库和 PyQuery 解析库的使用方法。

（2）学会使用 Selenium 基本库和 PyQuery 解析库实现淘宝网站信息的爬取。

知识准备

1. Selenium

Selenium 主要是用来做自动化测试的，支持多种浏览器，在爬虫中主要用来解决 JavaScript 渲染问题。在模拟浏览器进行网页加载时，当 urllib.requests 无法正常获取网页内容的时候，就可以尝试使用 Selenium 来爬取信息。

（1）声明浏览器对象。

```
#Python 文件名或者包名不要命名为 Selenium，否则会无法导入
from selenium import webdriver
#webdriver 可理解为浏览器的驱动器，其支持多种浏览器，这里以 Chrome 为例
browser = webdriver.Chrome()
```

（2）访问页面并获取网页 HTML。

```
from selenium import webdriver
```

```
browser = webdriver.Chrome()
#browser.page_Source 用于获取网页的全部 HTML
browser.get('HTTPS://www.taobao.com')
print(browser.page_Source)
browser.close()
```

（3）查找元素。

```
#查找单个元素
from selenium import webdriver
browser = webdriver.Chrome()
browser.get('HTTPS://www.taobao.com')
input_first = browser.find_element_by_id('q')
input_second = browser.find_element_by_css_selector('#q')
input_third = browser.find_element_by_xpath('//*[@id="q"]')print(input_first,input_
second,input_third)
browser.close()

#常用的查找方法
find_element_by_name
find_element_by_xpath
find_element_by_link_text
find_element_by_partial_link_text
find_element_by_tag_name
find_element_by_class_name
find_element_by_css_selector

#也可以使用通用方法进行查找
from selenium import webdriver
from selenium.webdriver.common.by import By
browser = webdriver.Chrome()
browser.get('HTTPS://www.taobao.com')
#第一个参数传入名称，第二个传入具体的参数
input_first = browser.find_element(BY.ID,'q')
```

```
print(input_first)
browser.close()


#查找多个元素
input_first = browser.find_elements_by_id('q')
```

（4）元素交互操作，即在搜索框中传入关键词并进行自动搜索。

```
from selenium import webdriver
import time
browser = webdriver.Chrome()
browser.get('HTTPS://www.taobao.com')
#找到搜索框
input = browser.find_element_by_id('q')
#传入关键词
input.send_keys('iPhone')
time.sleep(5)
#清空搜索框
input.clear()
input.send_keys('自行车')
#找到搜索按钮
button = browser.find_element_by_class_name('btn-search')
button.click()
```

（5）交互动作，即驱动浏览器进行动作，模拟拖动动作，将动作附加到动作链中并串行执行。

```
from selenium import webdriver
from selenium.webdriver import ActionChains
#引入动作链
browser = webdriver.Chrome()
url = 'HTTP://www.runoob.com/try/try.php?filename=jqueryui-api-droppable'
browser.get(url)
browser.switch_to.frame('iframeResult')
#切换到 iframeResult 框架
Source = browser.find_element_by_css_selector('#draggable')
```

```
#找到被拖动对象
target = browser.find_element_by_css_selector('#droppable')
#找到目标
actions = ActionChains(browser)
#声明 actions 对象
actions.drag_and_drop(Source, target)
#执行动作
actions.perform()
```

（6）执行 JavaScript。有些动作没有提供 API，如进度条下拉，其可以通过执行 JavaScript 来实现。

```
from selenium import webdriver
browser = webdriver.Chrome()
browser.get('HTTPS://www.zhihu.com/explore')
browser.execute_script('window.scrollTo(0, document.body.scrollHeight)')
browser.execute_script('alert("To Bottom")')
```

（7）获取元素信息。

```
#获取属性
from selenium import webdriver
from selenium.webdriver import ActionChains
browser = webdriver.Chrome()
url = 'HTTPS://www.zhihu.com/explore'
browser.get(url)
#获取网站 Logo
logo = browser.find_element_by_id('zh-top-link-logo')
print(logo)
print(logo.get_attribute('class'))
browser.close()

#获取文本值
from selenium import webdriver
browser = webdriver.Chrome()
url = 'HTTPS://www.zhihu.com/explore'
```

```
browser.get(url)
#input.text 文本值
input = browser.find_element_by_class_name('zu-top-add-question')print(input.text)
browser.close()
#获取 id、位置、标签名、大小
from selenium import webdriver
browser = webdriver.Chrome()
url = 'HTTPS://www.zhihu.com/explore'
browser.get(url)
input = browser.find_element_by_class_name('zu-top-add-question')
print(input.id)   #获取 id
print(input.location)   #获取位置
print(input.tag_name)   #获取标签名
print(input.size)   #获取大小
browser.close()
```

（8）Frame 操作。Frame 相当于独立的网页，如果在父类网页 Frame 中查找子类，则必须切换到子类的 Frame，如果子类需要查找父类，则也需要先切换到父类的 Frame。

```
from selenium import webdriver
from selenium.common.exceptions import NoSuchElementException
browser = webdriver.Chrome()
url = 'HTTP://www.runoob.com/try/try.php?filename=jqueryui-api-droppable'
browser.get(url)
browser.switch_to.frame('iframeResult')
Source = browser.find_element_by_css_selector('#draggable')
print(Source)
try:
logo = browser.find_element_by_class_name('logo')
except NoSuchElementException:
    print('NO LOGO')
browser.switch_to.parent_frame()
logo = browser.find_element_by_class_name('logo')
```

```
print(logo)
print(logo.text)
```

（9）等待。

① 隐式等待：当使用隐式等待执行的时候，如果 WebDriver 没有在 DOM 中找到元素，则继续等待，超出设定时间后将抛出找不到元素的异常，默认的时间是 0。

```
from selenium import webdriver
browser = webdriver.Chrome()
browser.implicitly_wait(10)
#10s 内无法加载出来就会抛出异常，10s 内加载出来则正常返回
browser.get('HTTPS://www.zhihu.com/explore')
input = browser.find_element_by_class_name('zu-top-add-question')
print(input)
```

② 显式等待：指定一个等待条件和一个最长等待时间，程序会判断在等待时间内条件是否满足，如果满足则返回，如果不满足则会继续等待，超过时间会抛出异常。

```
from selenium import webdriver
from selenium.webdriver.common.by import By
from selenium.webdriver.support.ui import WebDriverWait
from selenium.webdriver.support import expected_conditions as EC

browser = webdriver.Chrome()
browser.get('HTTPS://www.taobao.com/')
wait = WebDriverWait(browser, 10)
input = wait.until(EC.presence_of_element_located((By.ID, 'q')))
button = wait.until(EC.element_to_be_clickable((By.CSS_SELECTOR, '.btn-search')))
print(input, button)
```

（10）前进后退。实现浏览器的前进及后退，以浏览不同的网页。

```
import timefrom selenium import webdriver
browser = webdriver.Chrome()
browser.get('HTTPS://www.baidu.com/')
browser.get('HTTPS://www.taobao.com/')
browser.get('HTTPS://www.python.org/')
```

```
browser.back()
time.sleep(1)
browser.forward()
browser.close()
```

（11）Cookies 的设置。

```
from selenium import webdriver
browser = webdriver.Chrome()
browser.get('HTTPS://www.zhihu.com/explore')
print(browser.get_cookies())
browser.add_cookie({'name': 'name', 'domain': 'www.zhihu.com', 'value': 'germey'})
print(browser.get_cookies())
browser.delete_all_cookies()
print(browser.get_cookies())
```

（12）选项卡管理，增加浏览器窗口。

```
import time
from selenium import webdriver


browser = webdriver.Chrome()
browser.get('HTTPS://www.baidu.com')
browser.execute_script('window.open()')
print(browser.window_handles)
browser.switch_to_window(browser.window_handles[1])
browser.get('HTTPS://www.taobao.com')
time.sleep(1)
browser.switch_to_window(browser.window_handles[0])
browser.get('HTTP://www.fishc.com')
```

（13）异常处理。

```
from selenium import webdriver


browser = webdriver.Chrome()
browser.get('HTTPS://www.baidu.com')
browser.find_element_by_id('hello')
```

```
from selenium import webdriver
from selenium.common.exceptions import TimeoutException, NoSuchElementException

browser = webdriver.Chrome()
try:
browser.get('HTTPS://www.baidu.com')
except TimeoutException:
print('Time Out')
try:
browser.find_element_by_id('hello')
except NoSuchElementException:
print('No Element')
finally:
    browser.close()
```

2. PyQuery

PyQuery 相当于 jQuery 的 Python 实现，可以用于解析 HTML 网页等。它的语法与 jQuery 几乎完全相同。

（1）初始化。

PyQuery 有 4 种进行初始化的方法，可以通过传入“字符串”“lxml”“URL”“文件”来使用 PyQuery。

```
from pyquery import PyQuery as pq
from lxml import etree


doc = pq("<html></html>")  #传入字符串
doc = pq(etree.fromstring("<html></html>"))  #传入 lxml
doc = pq(url='HTTP://google.com/')  #传入 URL
doc = pq(filename=path_to_html_file)  #传入文件
```

现在，doc 就像 jQuery 中的$一样了。

下面通过一个简单的例子快速熟悉 PyQuery 的用法，传入文件 test.html，内容如下。

```
<div>
<tr class="c_1">
```

```
<td>Math</td>
<td>100</td>
</tr>
<tr class="c_2">
<td>English</td>
<td>100</td>
</tr>
</div>
```

使用 PyQuery，获取相关元素，代码如下。

```
from pyquery import PyQuery as pq   #导入 PyQuery
doc = pq(filename='test.html')   #传入文件 test.html
print doc.html()   #html()方法用于获取当前选中的 HTML 块
print doc('.c_1')   #相当于 class 选择器，选择 class 为 c_1 的 HTML 块

data = doc('tr')   #选择 tr 元素
for tr in data.items():   #遍历 data 中的 tr 元素
temp = tr('td').eq(1).text()   #选择第 2 个 td 元素中的文本块
print temp
```

其运行结果如下。

```
# print doc.html()
<div>
<tr class="c_1">
<td>Math</td>
<td>100</td>
</tr>
<tr class="c_2">
<td>English</td>
<td>100</td>
</tr>
</div>
# print doc('.c_1')
<tr class="c_1">
```

```
<td>Math</td>
<td>100</td>
</tr>

# print tr('td').eq(1).text()
100
# print tr('td').eq(2).text()
100
```

（2）PyQuery 获取元素的方法。

① html()和.text()：用于获取相应的 HTML 块或文本内容。

```
from pyquery import PyQuery as pq   #导入 PyQuery

doc=pq("<head><title>Hello World!</title></head>")  #传入字符串
print doc('head').html()   #获取相应的 HTML 块
print doc('head').text()   #获取相应的文本内容
```

其运行结果如下。

```
#print doc('head').html()
<title>hello</title>
#print doc('head').text()
Hello World!
```

② ('selector')：用于通过选择器来获取目标内容。

```
from pyquery import PyQuery as pq   #导入 PyQuery

doc = pq("<div><p id='id_1'>test1</p><p class='c_1'>test2</p></div>")

print doc('div').html()          #获取<div>元素内的 HTML 块
print doc('#id_1').text()        #获取 id 为 id_1 的元素内的文本内容
print doc('.c_1').text()         #获取 class 为 c_1 的元素的文本内容
```

其运行结果如下。

```
#print doc('div').html()
<p id='id_1'>test1</p><p class='c_1'>test2</p>
#print doc('#id_1').text()
```

```
test1
#print doc('.c_1').text()
test2
```

③ eq(index)：根据索引号获取指定元素（index 从 0 开始）。

```
from pyquery import PyQuery as pq   #导入 PyQuery


doc = pq("<div><p id='id_1'>test1</p><p class='c_1'>test2</p></div>")
print doc('p').eq(1).text()   #获取第二个 p 元素的文本内容
```

其运行结果如下。

```
#print doc('p').eq(1).text()
test2
```

④ find()：查找嵌套元素。

```
from pyquery import PyQuery as pq   #导入 PyQuery


doc = pq("<div><p id='id_1'>test1</p><p class='c_1'>test2</p></div>")
print doc('div').find('p')   #查找<div>内的 p 元素
#查找<div>内的 p 元素，输出第一个 p 元素
print doc('div').find('p').eq(0)
```

其运行结果如下。

```
#print doc('div').find('p')
<p id='id_1'>test1</p><p class='c_1'>test2</p>
#print doc('div').find('p').eq(0)
<p id='id_1'>test1</p>
```

⑤ filter()：根据 class、id 筛选指定元素。

```
from pyquery import PyQuery as pq   #导入 PyQuery


doc = pq("<div><p id='id_1'>test1</p><p class='c_1'>test2</p></div>")
print doc('p').filter('.id_1')   #查找 class 为 id_1 的 p 元素
print doc('p').filter('#c_1')   #查找 id 为 c_1 的 p 元素
```

其运行结果如下。

```
#print doc('p').filter('.id_1')
<p id='id_1'>test1</p>
```

```
#print doc('p').filter('#c_1')
<p class='c_1'>test2</p>
```

⑥ attr()：获取、修改属性值。

```
from pyquery import PyQuery as pq   #导入 PyQuery
doc = pq("<div><p id='id_1'>test1</p><p class='c_1'>test2</p></div>")

print doc('p').attr('id')                  #获取 p 标签的属性 id
print doc('p').attr('class','new')         #修改 a 标签的 class 属性为 new
```

其运行结果如下。

```
#print d('p').attr('id')
id_1
#print d('a').attr('class','new')
<p class="new">test2</p>
```

⑦ 其他操作。

addClass(value)：添加 class。

hasClass(name)：判断是否包含指定的 class，返回 True 或 False。

children()：获取子元素。

parents()：获取父元素。

next()：获取下一个元素。

nextAll()：获取后面全部的元素块。

not_('selector')：获取所有不匹配该选择器的元素。

for i in doc.items('li'):print i.text()：遍历 doc 中的 li 元素。

任务实施

抓取目标网页中商品的基本信息，包括商品图片、名称、价格、购买人数等。

（1）进入淘宝网页面，如图 2-15 所示。

页面下方有一个分页导航，如图 2-16 所示。

从中可以看出，这里商品的搜索结果最多为 100 页，想获取每一页的内容，只要将页码从 1 到 100 顺序遍历即可。直接在页面跳转文本框中输入要跳转的页码并单击“确定”按钮，即可进入相应的页面。

图 2-15　淘宝网页面

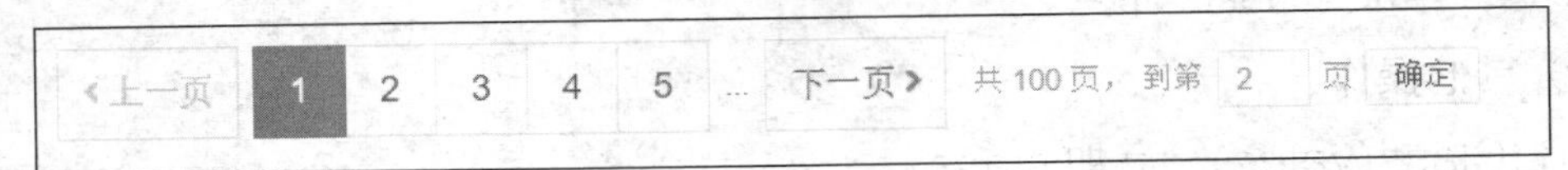

图 2-16　分页导航

（2）获取商品列表。

首先，导入相关的 Selenium 模块，代码如下。

```
from urllib.parse import quote
from selenium import webdriver
from selenium.webdriver.common.by import By
from selenium.common.exceptions import TimeoutException
from selenium.webdriver.support.wait import WebDriverWait
from selenium.webdriver.support import expected_conditions as EC
```

其次，构建一个 webdriver 对象，使用的浏览器为 Chrome。构建一个 WebDriverWait 对象，它指定了一个等待的最长时间，这里指定的最长时间为 10s，指定一个关键字“iPad”，代码如下。

```
browser = webdriver.Chrome()
wait = WebDriverWait(browser, 10)
KEYWORD = 'iPad'
```

定义一个 index_page()方法，用于抓取商品列表页。构造一个抓取的 URL：HTTPS://s.taobao.com/search?q=iPad。其中，参数 q 就是要搜索的关键字，在此将此关键字定义为一个变量。

在该方法中，先访问搜索商品的超链接，再判断当前的页码，如果页码大于 1，则进行跳页操作，否则等待页面加载完成。

在爬取到的页面中按“F12”键，进入开发者工具，查看相关节点信息。获取跳转文本框的 CSS 选择器，按快捷键“Ctrl+Shift+C”，单击跳转文本框，可发现跳转文本框对应的 HTML 代码高亮显示，选中代码并右键单击，在弹出的快捷菜单中，选择“Copy”中的“Copy selector”命令，如图 2-17 所示。

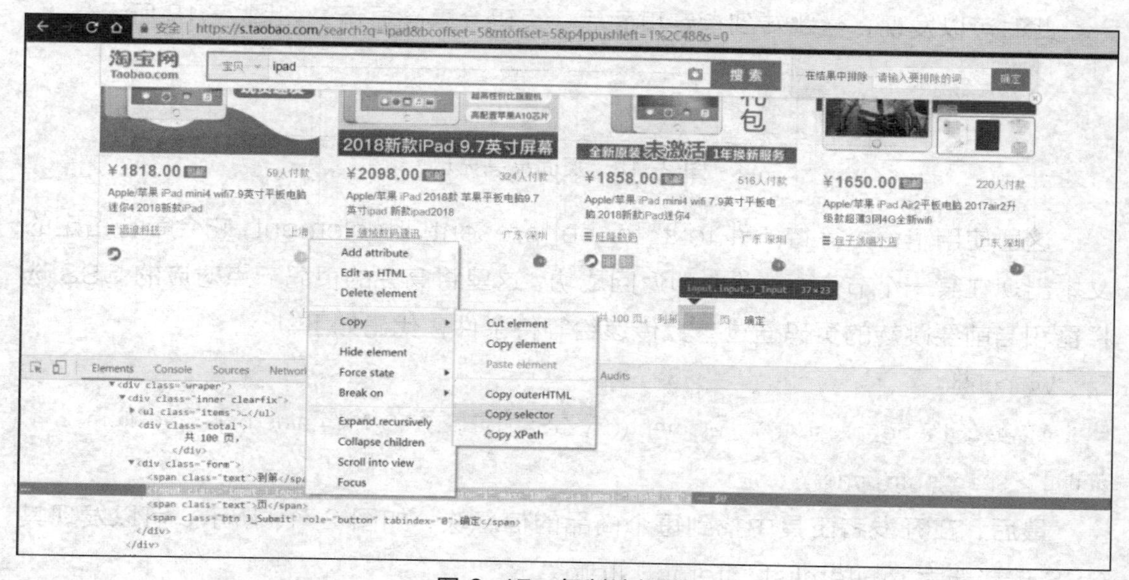

图 2-17　复制选择器

将查找到的元素赋值为 input。使用相同的方法获取“确定”按钮，并赋值为 submit，代码如下。

```
def index_page(page):
    print('正在爬取第', page, '页')
    try:
        url = 'https://s.taobao.com/search?q=' + quote(KEYWORD)
        browser.get(url)
        if page > 1:
            input = wait.until(
                EC.presence_of_element_located((By.CSS_SELECTOR, '#mainsrp-
pager div.form > input')))
            submit = wait.until(
```

```
            EC.element_to_be_clickable((By.CSS_SELECTOR, '#mainsrp-pager
div.form > span.btn.J_Submit')))
```

当页码大于 1 时，清空跳转文本框，再次调用 send_keys()方法填充页码，单击“确定”按钮即可，代码如下。

```
    input.clear()
    input.send_keys(page)
    submit.click()
```

此时可以发现，当跳转到指定页面后，页码会高亮显示，如图 2-18 所示。

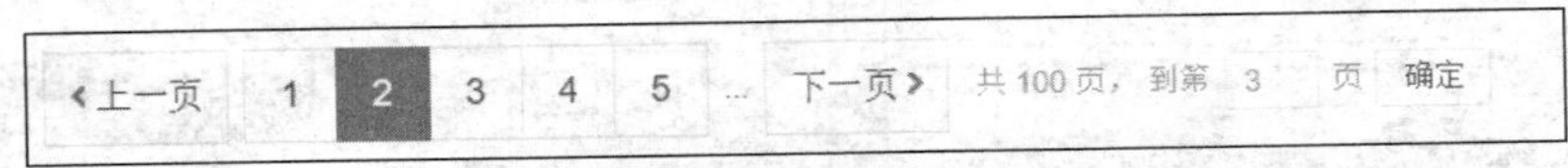

图 2-18　页面高亮显示

这里使用了一个等待条件 text_to_be_present_in_element，它会等待指定的文本出现在某一个节点中，此时即返回成功。这里将高亮的页码节点对应的 CSS 选择器和当前要跳转的页码通过参数传递给等待条件，代码如下。

```
    wait.until(
    EC.text_to_be_present_in_element((By.CSS_SELECTOR,'#mainsrp-pagerli.item.
active > span'), str(page)))
```

最后，在开发者工具中找到每个商品的信息块，如图 2-19 所示，可以发现其 CSS 选择器为 m-itemlist、items、item。

```
▼<div id="mainsrp-itemlist"> == $0
  ▼<div class="m-itemlist" data-spm="14">
    ▼<div class="grid g-clearfix">
       ::before
     ▼<div class="items">
       ▶<div class="item J_MouserOnverReq item-ad  " data-category="auctions" data-index="0">…</div>
       ▶<div class="item J_MouserOnverReq item-ad  " data-category="auctions" data-index="1">…</div>
       ▶<div class="item J_MouserOnverReq  " data-category="auctions" data-index="2">…</div>
       ▶<div class="item J_MouserOnverReq  " data-category="auctions" data-index="3">…</div>
       ▶<div class="item J_MouserOnverReq  " data-category="auctions" data-index="4">…</div>
       ▶<div class="item J_MouserOnverReq  " data-category="auctions" data-index="5">…</div>
       ▶<div class="item J_MouserOnverReq  " data-category="auctions" data-index="6">…</div>
       ▶<div class="item J_MouserOnverReq  " data-category="auctions" data-index="7">…</div>
       ▶<div class="item J_MouserOnverReq  " data-category="auctions" data-index="8">…</div>
       ▶<div class="item J_MouserOnverReq  " data-category="auctions" data-index="9">…</div>
       ▶<div class="item J_MouserOnverReq  " data-category="auctions" data-index="10">…</div>
       ▶<div class="item J_MouserOnverReq  " data-category="auctions" data-index="11">…</div>
       ▶<div class="item J_MouserOnverReq  " data-category="auctions" data-index="12">…</div>
       ▶<div class="item J_MouserOnverReq  " data-category="auctions" data-index="13">…</div>
       ▶<div class="item J_MouserOnverReq  " data-category="auctions" data-index="14">…</div>
       ▶<div class="item J_MouserOnverReq  " data-category="auctions" data-index="15">…</div>
       ▶<div class="item J_MouserOnverReq  " data-category="auctions" data-index="16">…</div>
       ▶<div class="item J_MouserOnverReq  " data-category="auctions" data-index="17">…</div>
       ▶<div class="item J_MouserOnverReq  " data-category="auctions" data-index="18">…</div>
       ▶<div class="item J_MouserOnverReq  " data-category="auctions" data-index="19">…</div>
```

图 2-19　每个商品的信息块

将 m-itemlist、items、item 传入 presence_of_element_located()方法中。如果加载成功，则会执行后面的 get_products()方法，代码如下。

```
    wait.until(EC.presence_of_element_located((By.CSS_SELECTOR, '.m-itemlist.
items.item')))
        get_products()
    except TimeoutException:
        index_page(page)
```

通过编写 get_products()方法来解析商品列表。这里可直接获取商品源代码，并使用 PyQuery 进行解析，代码如下。

```
from pyquery import PyQuery as pq
def get_products():
    html = browser.page_Source
    doc = pq(html)
```

此时，可发现每一个商品块都被封装在一个 class 为 J_MouserOnverReq 的 div 标签中，div 标签中存有商品的基本信息。使用 PyQuery 提供的方法将所需的信息提取出来并存储到一个字典中，代码如下。

```
items = doc('#mainsrp-itemlist.items.item').items()
for item in items:
    product = {
        'image': item.find('.pic .img').attr('data-src'),
        'price': item.find('.price').text(),
        'deal': item.find('.deal-cnt').text(),
        'title': item.find('.title').text(),
        'shop': item.find('.shop').text(),
        'location': item.find('.location').text()
    }
```

输出提取的信息，并将这些信息传入 save_to_mongo 方法，代码如下。

```
print(product)
save_to_mongo(product)
```

（3）保存到 MongoDB。

将商品信息保存到 MongoDB 中。

首先，导入 PyMongo 模块。

其次，创建一个 MongoDB 的连接对象，并指定数据库和集合的名称，代码如下。

```
import pymongo
MONGO_URL = 'localhost'
MONGO_DB = 'taobao'
MONGO_COLLECTION = 'products'
client = pymongo.MongoClient(MONGO_URL)
db = client[MONGO_DB]
```

最后，创建 save_to_mongo 方法，调用 insert()方法将数据插入 MongoDB，代码如下。

```
def save_to_mongo(result):
    try:
        if db[MONGO_COLLECTION].insert(result):
            print('存储到 MongoDB 成功')
    except Exception:
        print('存储到 MongoDB 失败')
```

此处的 result 变量就是 get_products()方法中传入的 product，包含单个商品信息。

（4）遍历每页。

定义的 get_index 方法中需要接收参数 page（代表页码），在这里将实现页码的遍历，最后需要将 browser 关闭，代码如下。

```
MAX_PAGE = 100
def main():
    for i in range(1, MAX_PAGE + 1):
        index_page(i)
    browser.close()
```

（5）运行。

运行代码，将发现其首先会打开一个 Chrome 浏览器窗口，并会访问淘宝网页面，输入框中会自动填入搜索关键字，控制台会输出相应的结果，如图 2-20 所示。

```
正在爬取第 1 页
{'image': '//g-search1.alicdn.com/img/bao/uploaded/i4/imgextra/i3/13022581/TB2._5ckYZnBKNjSZFKXXcGOVXa_!!0-saturn_solar.jpg_360x360Q90.jpg',
 'price': '1999.00', ' deal': '258人付款', ' title': '送电动牙刷【2年保修】Apple/苹果 Ipad 9.7英寸WLAN平板电脑2017款', 'shop': '绿森数码官方旗舰店
 ', 'location': '浙江 杭州'}
存储到MongoDB成功
{'image': '//g-search3.alicdn.com/img/bao/uploaded/i4/i3/1669409267/TB1CACiGL5TBuNjSspmXXaDRVXa_!!0-item_pic.jpg_360x360Q90.jpg', 'price':
 '2568.00', ' deal': '1474人付款', ' title': '[12期分期]Apple/苹果 iPad mini 4 7.9英寸 平板电脑 128GWifi版', 'shop': '卓辰数码旗舰店',
 'location': '浙江 杭州'}
存储到MongoDB成功
{'image': '//g-search1.alicdn.com/img/bao/uploaded/i4/i1/197232874/TB132Qat98YBeNkSnb4XXaevFXa_!!0-item_pic.jpg_360x360Q90.jpg', 'price':
 '2299.00', ' deal': '5000人付款', ' title': '赠电动牙刷【2年保修】Apple/苹果 iPad 2018款 9.7英寸平板电脑', 'shop': '绿森数码官方旗舰店',
 'location': '浙江 杭州'}
存储到MongoDB成功
{'image': '//g-search3.alicdn.com/img/bao/uploaded/i4/i1/97045700/TB2qoUdyuuSBuNjSsplXXbe8pXa_!!97045700.jpg_360x360Q90.jpg', 'price':
 '1878.00', ' deal': '3791人付款', ' title': 'Apple/苹果 iPad 2018款 苹果平板电脑9.7英寸ipad 新款ipad2018', 'shop': '深圳_恒波', 'location':
 '广东 深圳'}
存储到MongoDB成功
```

图 2-20　控制台输出的相应结果

任务 3　使用 Scrapy 爬取北京公交信息

任务描述

（1）借助学习论坛、网络视频等网络资源和各种图书资源，学习网络爬虫相关技术，熟悉 Scrapy 的使用。

（2）使用 Scrapy 实现淘宝网站信息的获取。

任务目标

学会使用 Scrapy 爬取北京公交信息。

知识准备

Scrapy 是一个为了爬取网站数据、提取结构性数据而编写的应用框架，可以应用在包括实现数据挖掘功能、信息处理功能或存储历史数据功能等的一系列处理程序中。

Scrapy 的操作流程如下。

1. 选择一个网站

当需要从某个网站中获取信息，但该网站未提供 API 或获取信息的机制时，使用 Scrapy 将可以轻松实现信息的获取。

选择一个网站，确定爬取目标的 URL，并对其进行初步分析，确定需要爬取的

信息。

2. 创建一个 Scrapy 项目

在开始爬取之前，必须创建一个新的 Scrapy 项目。切换路径到项目计划保存目录，并执行以下命令。

```
scrapy startproject work
```

该命令将会创建包含下列内容的 work 目录。

```
work/
    scrapy.cfg
    work/
        __init__.py
    items.py
    pipelines.py
    settings.py
    spiders/
    __init__.py
            ...
```

scrapy.cfg：项目的配置文件。

work/：该项目的 Python 模块，之后将在此创建爬虫程序。

items.py：项目中的 item 文件。

pipelines.py：项目中的 pipelines 文件。

settings.py：项目的设置文件。

spiders/：放置 spider 代码的目录。

3. 创建一个 spider

切换目录到“work”，并使用 genspider 语句创建一个 spider。语句的格式如下。

```
scrapy genspider spider_name（自行定义 spider 的名称） URL（目标网址）
```

4. 定义 Item

Item 是保存爬取到的数据的容器；其使用方法和 Python 字典类似，并且提供了额外的保护机制来避免拼写错误导致的未定义字段错误。

根据从目标网站获取到的数据对 Item 进行建模，并在 Item 中定义相应的字段。编辑 work 目录中的 items.py 文件，格式类似于：

```
import scrapy
```

```
class myItem(scrapy.Item):
title = scrapy.Field()
link = scrapy.Field()
desc = scrapy.Field()
```

通过定义 Item，可以很方便地使用 Scrapy 的其他方法，而这些方法需要知道 Item 的定义。

5. 编写 spider

spider 是用户编写的用于从单个网站（或者一些网站）爬取数据的类。其包含了下载初始 URL、获取网页中超链接，以及提取页面中的内容生成 item 的方法。

为了创建一个 spider，必须继承 scrapy.Spider 类，且定义以下 3 个属性。

（1）name：用于区分 spider。该名称必须是唯一的，不可以为不同的 spider 设定相同的名称。

（2）start_urls：包含了 spider 在启动时进行爬取的 URL 列表。因此，第一个被获取到的页面将是其中之一。后续的 URL 则从初始的 URL 获取到的数据中提取。

（3）parse()：spider 的一个方法。被调用时，每个初始 URL 完成下载后生成的 Response 对象将会作为唯一的参数传递给该函数。该方法负责解析返回的数据，提取数据（生成 Item）以及生成需要进一步处理的 URL 的 Request 对象。

6. 提取 Item

从网页中提取数据有很多方法。Scrapy 使用了一种基于 Xpath 和 CSS 表达式的机制：Scrapy Selectors。这里给出常用的 Xpath 表达式以及对应的含义。

/html/head/title：选择 HTML 文档中 head 标签内的 title 元素。

/html/head/title/text()：选择前面提到的 title 元素的文字。

//td：选择所有的 td 元素。

//div[@class="mine"]：选择所有具有 class="mine"属性的 div 元素。

为了配合 Xpath，Scrapy 除了提供了 selector 之外，还提供了方法来避免每次从 Response 中提取数据时生成 selector 的麻烦。

selector 有以下 4 个基本的方法（单击相应的方法可以看到详细的 API 文档）。

xpath()：传入 Xpath 表达式，返回该表达式所对应的所有节点的 selector 列表。

css()：传入 CSS 表达式，返回该表达式所对应的所有节点的 selector 列表。

extract()：序列化该节点为 Unicode 字符串并返回 list。

re()：根据传入的正则表达式对数据进行提取，返回 Unicode 字符串的 list。

7. 存储爬取的数据

最简单的存储爬取数据的方式是使用 Feed exports。

```
scrapy crawl work -o items.json
```

该命令将采用 JSON 格式对爬取的数据进行序列化，生成 items.json 文件。

对于小规模的项目，这种存储方式较为灵活。如果需要对爬取到的 Item 做更多、更复杂的操作，则需要编写 pipelines.py。

8. 执行项目

切换到项目的根目录，执行下列命令启动 spider。

```
scrapy crawl work
```

Scrapy 为 spider 的 start_urls 属性中的每个 URL 都创建了 scrapy.Request 对象，并将 parse()方法作为回调函数赋值给了 Request 对象。

Request 对象经过调度生成 scrapy.HTTP.Response 对象并送回给 spider parse()方法。

任务实施

（1）在命令行窗口中新建一个名为 beibus 的 Scrapy 项目，如图 2-21 所示。

```
C:\Users\mac>scrapy startproject beibus
New Scrapy project 'beibus', using template directory 'c:\\users\\mac\\appdata\\local\\programs\\python\\python36\\lib\\
site-packages\\scrapy\\templates\\project', created in:
    C:\Users\mac\beibus

You can start your first spider with:
    cd beibus
    scrapy genspider example example.com
```

图 2-21　创建 Scrapy 项目

（2）切换到 beibus 目录，并使用 genspider 创建一个 spider，如图 2-22 所示。

```
C:\Users\mac>cd beibus

C:\Users\mac\beibus>scrapy genspider bei_bus beijing.8684.cn
Created spider 'bei_bus' using template 'basic' in module:
  beibus.spiders.bei_bus
```

图 2-22　创建 spider

（3）使用 PyCharm 打开 beibus 项目，其结构如图 2-23 所示。

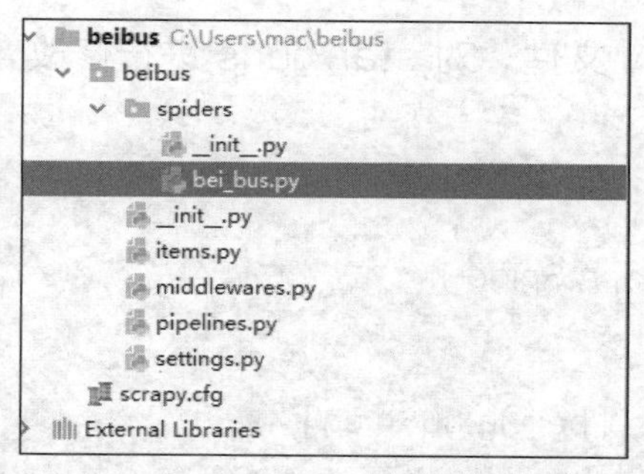

图 2-23　beibus 项目的结构

（4）进入 settings，将 ROBOTSTXT_OBEY 的参数改为 False，使爬虫不遵守 Robots 协议，代码如下。

```
ROBOTSTXT_OBEY = False
```

（5）将 DEFAULT_REQUEST_HEADERS 方法的注释去掉，并在其中添加 User-Agent 属性，代码如下。

```
DEFAULT_REQUEST_HEADERS = {
'Accept': 'text/html,application/xhtml+xml,application/xml;q=0.9,*/*;q=0.8',
 'Accept-Language': 'en',
'User-Agent': 'Mozilla/5. 0 (Windows NT 10.0; Win64; x64) AppleWebKit/537.36
|(KHTML, like Gecko) Chrome/67. 0. 3396. 99 Safari/537. 36',
```

在浏览器页面中右键单击，在弹出的快捷菜单中选择“检查”命令，在“Network”标签的“Request Headers”中可以找到 User-Agent 属性，如图 2-24 所示。

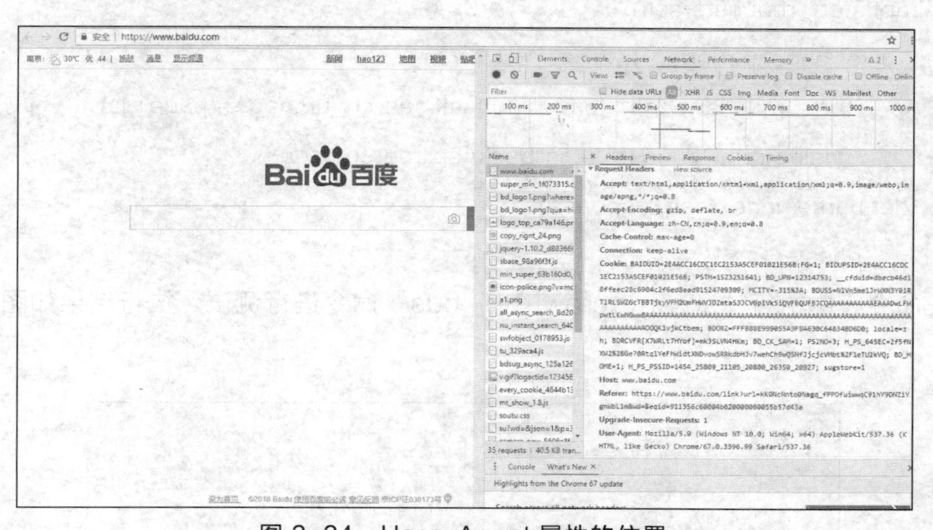

图 2-24　User-Agent 属性的位置

此操作的功能是使爬虫实现模拟浏览器访问的效果。

（6）进入 bei_bus.py 文件，对 start_urls 进行修改，代码如下。

```
import scrapy


class BeiBusSpider(scrapy.Spider):
    name = 'bei_bus'
    allowed.domains = ['beijing.8684.cn']
    search_url = 'http://beijing.8684.cn'


  def parse(self,response)
     pass
```

（7）新建一个 start_requests()方法（此方法名固定），获得并构建一级网页 URL，代码如下。

```
from scrapy import Spider,FormRequest,Request


class BeiBusSpider(scrapy.Spider):
    name = 'bei_bus'
    allowed.domains = ['beijing.8684.cn']
    search_url = 'http://beijing.8684.cn'


  def start_requests(self):
     for page in range(9):
        url= '{url}/list{page}'.format(url=self.search_url,page=(page+1))
        yield FormRequest(url,callback=self.parse_index)
  def parse_index(self,response):
      pass
```

在命令行中执行“scrapy crawl bei_bus”命令进行测试，执行结果如图 2-25 所示。

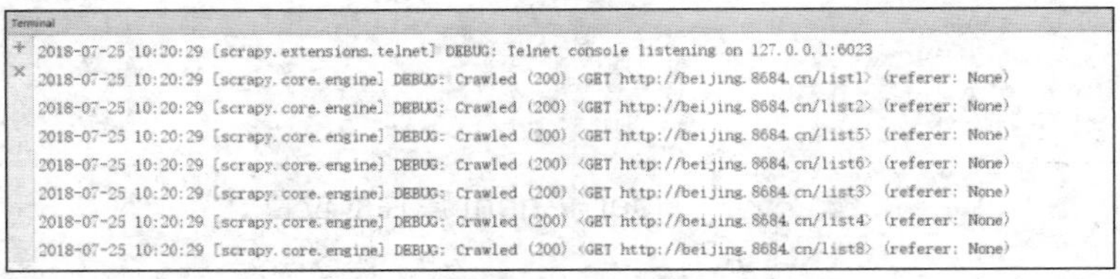

```
Terminal
2018-07-25 10:20:29 [scrapy.extensions.telnet] DEBUG: Telnet console listening on 127.0.0.1:6023
2018-07-25 10:20:29 [scrapy.core.engine] DEBUG: Crawled (200) <GET http://beijing.8684.cn/list1> (referer: None)
2018-07-25 10:20:29 [scrapy.core.engine] DEBUG: Crawled (200) <GET http://beijing.8684.cn/list2> (referer: None)
2018-07-25 10:20:29 [scrapy.core.engine] DEBUG: Crawled (200) <GET http://beijing.8684.cn/list5> (referer: None)
2018-07-25 10:20:29 [scrapy.core.engine] DEBUG: Crawled (200) <GET http://beijing.8684.cn/list6> (referer: None)
2018-07-25 10:20:29 [scrapy.core.engine] DEBUG: Crawled (200) <GET http://beijing.8684.cn/list3> (referer: None)
2018-07-25 10:20:29 [scrapy.core.engine] DEBUG: Crawled (200) <GET http://beijing.8684.cn/list4> (referer: None)
2018-07-25 10:20:29 [scrapy.core.engine] DEBUG: Crawled (200) <GET http://beijing.8684.cn/list8> (referer: None)
```

图 2-25　获得并构建一级网页 URL

start_requests()方法使用 format 拼接 URL，并用 yield 再次发起网页访问，callback 后的方法为回调函数，可以在回调函数中对网页进行处理。

（8）在 parse_index()方法中，使用 Xpath 寻找详细信息页面的 URL，并对网址进行拼接，之后对二级网页发起请求，并返回回调函数 parse_detail()，代码如下。

```
from scrapy import Spider, FormRequest, Request
from urllib.parse import urljoin


class BeiBusSpider(Spider):
    name = 'bei_bus'
    allowed.domains = ['beijing.8684.cn ']
    search_url = 'http://beijing.8684.cn'


    def start_requests(self):
        for page in range(9):
            url = '{url}/list {page}'.format(url=self.search_url, page= (page+1))
            yield FormRequest(url, callback=self.parse_index)
    def parse_index(self, response):
        beijingbus=response.xpath('//*[@id="con_site_1"]/a//@href').extract()
        for href in beijingbus:
            url2 = urljoin(self.search_url, href)
            yield Request(url2, callback=self.parse_detail)


    def parse_detail(self, response):
        pass
```

extract()方法表示选择符合条件的所有文本，Xpath 中的参数表示搜索页面中 id 为 con_site_1 的 a 标签中的 href 属性，并使用 urljoin()方法拼接 URL。

（9）在 parse_detail()方法中对详细信息页面进行如下操作。

```
def parse_detail (self，response):
bus_name=response.xpath('//*[@id="bus_line"]/div[1]/div/div/hl//text()').extract_first()
bus_type=response.xpath('//*[@id="bus_line"]/div[1]/div/div/a//text()').extract_first()
bus_time=response.xpath('//*[@id="bus_line"]/div[1]/div/p[1]//text()').extract_first()
```

```
ticket = response.path('//*[@id="bus_line"]/div[1]/div/p[2]//text()').extract_first()
gongsi = response.xpath('//*[@id="bus_line"]/div[1]/div/p[3]/a//text()').extract_first()
gengxin = response.xpath('//*[@id="bus_line"]/div[1]/div/p[4]//text()').extract_first()
try:
    licheng = response.css('div p[class="bus_label_t2"]:: text').extract_first()
except:
    licheng = None
try:
    wang_info = response.path('//*[@id="bus_line"]/div[4]/div/strong//text()').extract_first()
except:
    wang_info = None
try:
    wang_list_tag = response.xpath('//*[@id="bus_line"]/div[5]/div/div/a//text()').extract()
except:
    wang_list_tag = None
try:
    fan_info = response.xpath('//*[@id="bus_line"]/div[6]/div/strong//text()').extract_first()
except:
    fan_info = None
try:
    fan_list_tag = response.xpath('//*[@id="bus_line"]/div[7]/div/div/a//text()').extract()
except:
    fan_list_tag = None
```

text()用于获取标签下的文本，extract_first()方法用于表示只返回符合条件的第一个数据。CSS 选择器中的语句表示选择 div 标签中包含 p 标签且 class 属性为 bus_label_t2 的数据。

（10）使用如下语句对 wang_list_tag 字段和 fan_list_tag 字段进行格式化。

```
fan_buff = ''
if wang_list_tag:
for wang in wang_list_tag:
    wang_buff += wang + ","
if fan_list_tag:
```

```
    for fan in fan_list_tag:
        fan_buff += fan+","
```

（11）进入 items.py 文件，修改其中的 class 以格式化数据，代码如下。

```
import scrapy
class BeibusItem(scrapy.Item):
    bus_name = scrapy.Field ()
    bus_type = scrapy.Field ()
    bus_time = scrapy.Field ()
    ticket = scrapy.Field ()
    gongsi = scrapy.Field ()
    gengxin = scrapy.Field ()
    licheng = scrapy.Field ()
    wang_info = scrapy.Field()
    wang_buff = scrapy.Field()
    fan_info = scrapy.Field ()
    fan_buff = scrapy.Field()
```

（12）在 bei_bus.py 文件的 parse_detail()方法的末尾添加如下语句，以格式化数据。

```
bus_item = BeibusItem()
for field in bus_item.fields:
    bus_item[field] = eval(field)
yield bus_item
```

（13）在 MySQL 的 studb 数据库中创建一个 stu_businfo 表，其格式如图 2-26 所示。

```
mysql> CREATE TABLE stu_businfo(
    -> bus_name varchar(1000),
    -> bus_type varchar(1000),
    -> bus_time varchar(1000),
    -> ticket varchar(1000),
    -> gongsi varchar(1000),
    -> gengxin varchar(1000),
    -> licheng varchar(1000),
    -> wang_info varchar(1000),
    -> wang_buff varchar(1000),
    -> fan_info varchar(1000),
    -> fan_buff varchar(1000))
    -> DEFAULT CHARSET=utf-8;
```

图 2-26　stu_businfo 表的格式

（14）在 settings.py 文件末尾添加如下参数。

```
DB_HOST = 'localhost'
DB_USER = 'stu'
DB_PWD = 'stu123'
DB_CHARSET='UTF8'
```

（15）在 pipelines.py 文件中，修改 class 名称为 MySQL Pipeline，添加初始化方法，将 host、user、password、db、charset 从 settings 中读取出来，并通过添加一个 connect()方法建立与数据库的连接，代码如下。

```
import pymysql
from beibus import settings


class MysqlPipeline(object):
    def __init__(self):
        self.host = settings.DB_HOST
        self.user = settings.DB_USER
        self.pwd = settings.DB_PWD
        self.db = settings.DB
        self.charset = setting.DB.CHARSET
        self.connect()

    def connect(self):
        self.conn = pymysql.connect(host=self.host,
                                    user=self.user,
                                    password=self.pwd,
                                    db=self.db,
                                    charset=self.charset)
        self.cursor = self.conn.cursor()
```

（16）添加一个 close_spider()方法，用于关闭 MySQL 数据库连接，代码如下。

```
def close_spider(self, spider)
    self.conn.close()
    self.cursor.close()
```

（17）实现 process_item()方法，用于完成向数据库中插入数据的操作，代码如下。

```
    def process_item(self,item,spider):
        sql ='INSERT INTO stu_businfo (bus_name, bus_type, bus_time, ticket, gongsi, gengxin, licheng, wang_info,wang_buff, fan_info, fan_buff) VALUES ("%s","%s","%s","%s","%s","%s","%s","%s","%s","%s","%s")'%(item['bus_name'], item['bus_type'],item['bus_time'], item['ticket'],item['gongsi'],item['gengxin'],item['licheng'],item['wang_info'],item['wang_buff'], item['fan_info'],item['fan_buff'])
        #执行 SQL 语句
        self.cursor.execute(sql)
        self.conn.commit()
        return item
```

（18）在 settings.py 中将 ITEM_PIPELINES 方法的注释去掉，并将其中的内容改为“'beibus.pipelines.MysqlPipeline':300”，后面的数字代表优先级，数字越小，优先级越高，代码如下。

```
ITEM_PIPELINES ={
    'beibus.pipelines.MysqlPipeline': 300
```

在 PyCharm 软件左下角单击“Terminal”标签，进入命令行界面，在其中执行“scrapy crawl bei_bus”命令，启动爬虫项目，如图 2-27 所示。

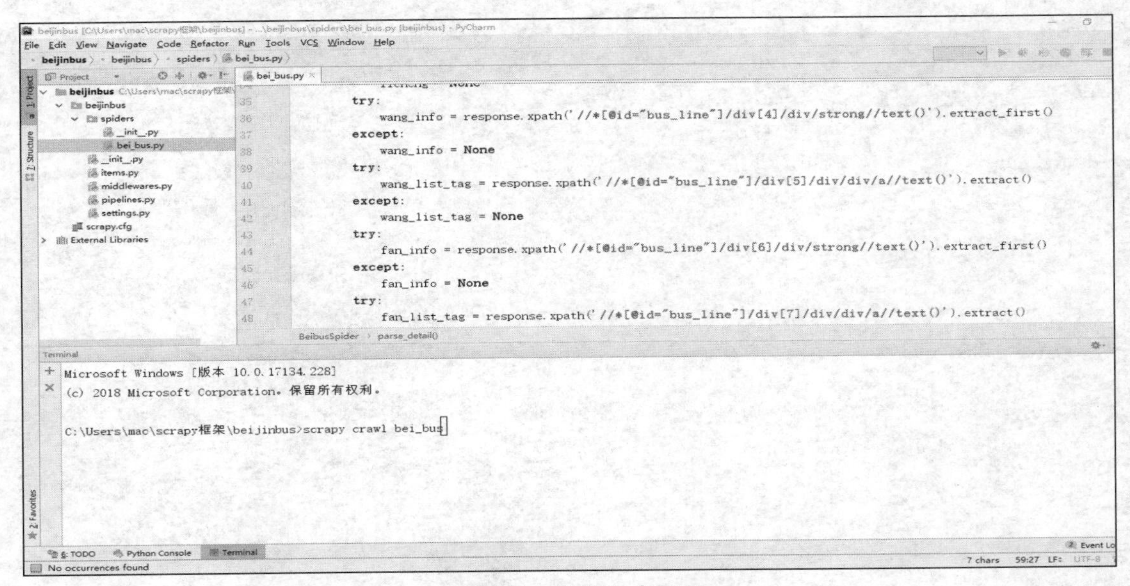

图 2-27　启动爬虫项目

任务 4 创新与拓展

任务描述

使用 Python 语言实现网络爬虫编程，爬取 https://news.163.com/网页中相应的信息。

（1）爬取页面中新闻的标题名称及相应的详细文字和视频地址信息等。

（2）实现多线程网络爬虫编程，爬取“图片”板块中的图片地址信息。

任务目标

（1）知道 urllib 基本库和 BeautifulSoup 解析库的使用方法。

（2）学会使用 urllib 基本库和 BeautifulSoup 解析库进行网页中相关信息的爬取。

项目 3
日志数据采集实践

学习目标

【知识目标】

① 了解 Flume 的特点，熟悉 Flume 的工作原理。
② 熟悉 Flume 运行的核心 Agent，识记 Source、Channel、Sink 的概念。

【技能目标】

① 学会 Flume 的安装和不同应用场景下的配置。
② 学会 Flume 采集数据上传到 HDFS 的方法。
③ 学会 Flume 采集数据上传到 HBase 的方法。

项目描述

许多公司的业务平台每天都会产生大量的日志数据。对这些日志数据进行采集，并进行数据分析，可以挖掘公司业务平台日志数据中的潜在价值，为公司决策和公司后台服务器平台性能评估提供可靠的数据保证。

目前常用的开源日志收集系统有很多。Apache Flume 是一个分布式、高可靠、高可用的服务，用于高效地收集、聚合和移动大量的日志数据，它具有基于流式数据流的简单灵活的架构。其可靠性机制和许多故障转移及恢复机制使 Flume 具有强大的容错能力。

本项目主要完成 Flume 的安装和配置，并按照大数据企业工作流程和规范，完成 Flume 数据的采集。

任务 1 Flume 的安装和配置

任务描述

（1）完成 Flume 相关基础知识的学习。

（2）完成 Flume 的安装。

（3）使用 Flume 采集数据的常用方式，即通过配置 Flume 的 Agent 信息，以及定义 Flume 的数据源、采集方式和输出目标，完成数据采集的关键参数配置工作。

任务目标

（1）熟悉 Flume 的相关基础知识。

（2）学会 Flume 的安装和不同应用场景下的配置。

知识准备

1. Flume

Flume 是一个分布式、高可靠和高可用的海量日志采集、聚合及传输服务。Flume 支持在日志系统中定制各类数据发送方，用于收集数据；同时，Flume 提供了对数据进行简单处理，并写到各种数据接收方（如文本、HDFS、HBase 等）的功能。其设计原理是将数据流（如日志数据）从各种网站服务器中汇集起来，并存储到 HDFS、HBase 等集中存储器中。Flume 数据采集的工作流程如图 3-1 所示。

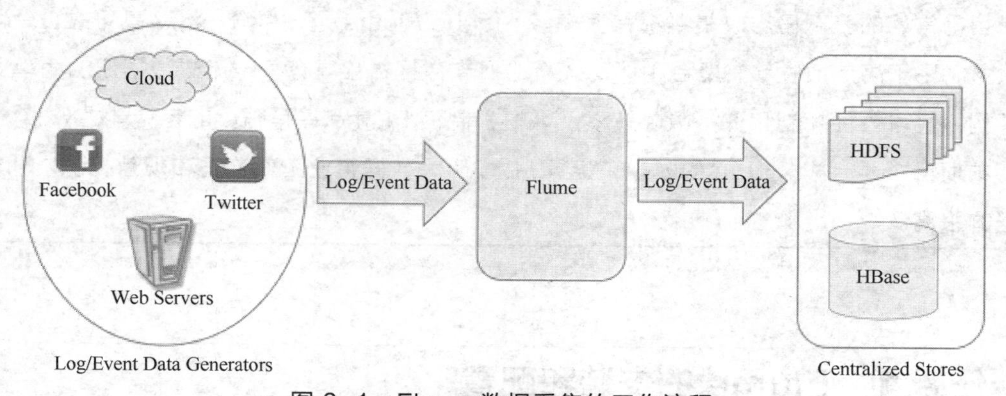

图 3-1　Flume 数据采集的工作流程

2. Flume 的工作原理

Flume 的数据流由事件（Event）贯穿始终。事件是 Flume 的基本数据单位，它携带日志数据（字节数组形式）并且携带有头信息，这些事件由 Agent 外部的 Source 生成，当 Source 捕获事件后会进行特定的格式化，把事件推入（单个或多个）Channel。可以把 Channel 看作一个缓冲区，它将保存事件直到 Sink 处理完该事件。Sink 负责持久化日志或者把事件推向另一个 Source。Flume 的外部结构如图 3-2 所示。

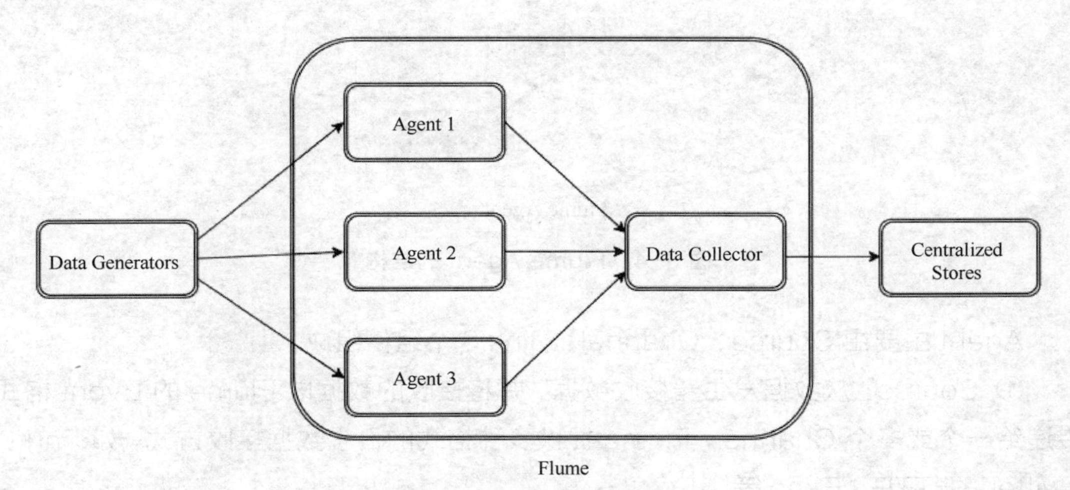

图 3-2　Flume 的外部结构

图 3-2 中的 Data Generators（数据发生器，如 Facebook、Twitter）产生的数据被单个 Agent 所收集（单个 Agent 部署在 Data Generator 上），之后 Data Collector（数据收集器）从各个 Agent 上汇集数据，并将采集到的数据存入 HDFS 或 HBase 中。

（1）Flume 事件

事件作为 Flume 内部数据传输的最基本单元，是由一个转载数据的字节数组（该数组从数据源接入点传入，并传输给传输器，即 HDFS、HBase）和一个可选头部构成的。典型的 Flume 事件的数据结构如图 3-3 所示。

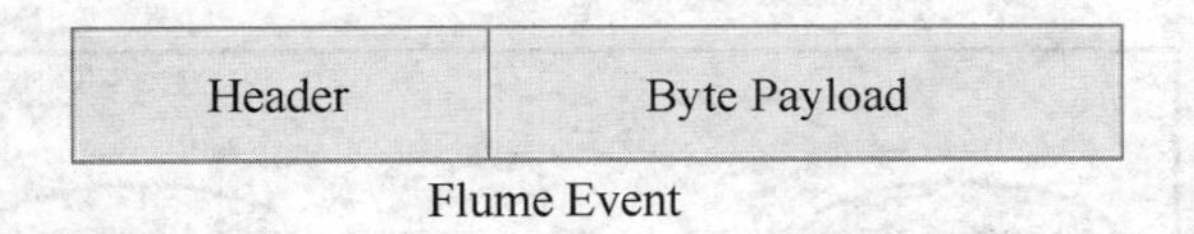

图 3-3　典型的 Flume 事件的数据结构

（2）Flume Agent

在了解了 Flume 的外部结构之后，可知 Flume 内部有一个或者多个 Agent，然而，对于每一个 Agent 来说，Agent 就是一个独立的守护进程（JVM），它从客户端接收数据，或者从其他的 Agent 上接收数据，并迅速地将获取的数据传给下一个目的节点 Sink 或 Agent。Flume Agent 的结构如图 3-4 所示。

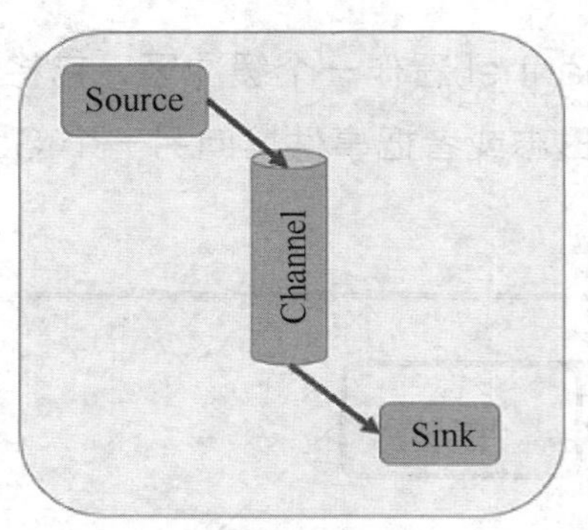

图 3-4　Flume Agent 的结构

Agent 主要由 Source、Channel、Sink 3 个组件组成。

① Source：从数据发生器接收数据，并将接收的数据以 Flume 的 Event 格式传递给一个或多个 Channel，Flume 提供多种格式的日志数据接收方式，如 Thrift、Twitter、HTTP、Exec 等。

② Channel：Channel 是一种短暂的存储容器，它将从 Source 处接收到的 Event 格式的数据缓存起来，直到它们被 Sink 消费，Channel 在 Source 和 Sink 间起到了桥梁的作用。Channel 是一个完整的事务，这一点保证了数据在收发时的一致性，并且它可以和任意数量的 Source 和 Sink 链接，其支持的类型有 JDBC Channel、File System Channel、Memory Channel 等。

③ Sink：Sink 将数据存储到集中存储器（如 HBase 和 HDFS）中，它从 Channel 接收数据并将其传递到目的地。目的地可能是另一个 Sink，也可能是 HDFS、HBase。

Agent 的常用组合形式如图 3-5 和图 3-6 所示。

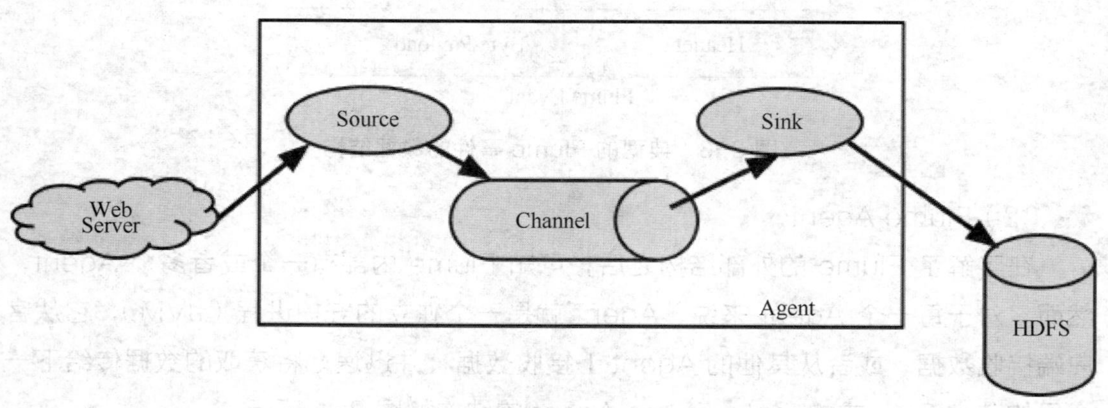

图 3-5　Agent 的常用组合形式（1）

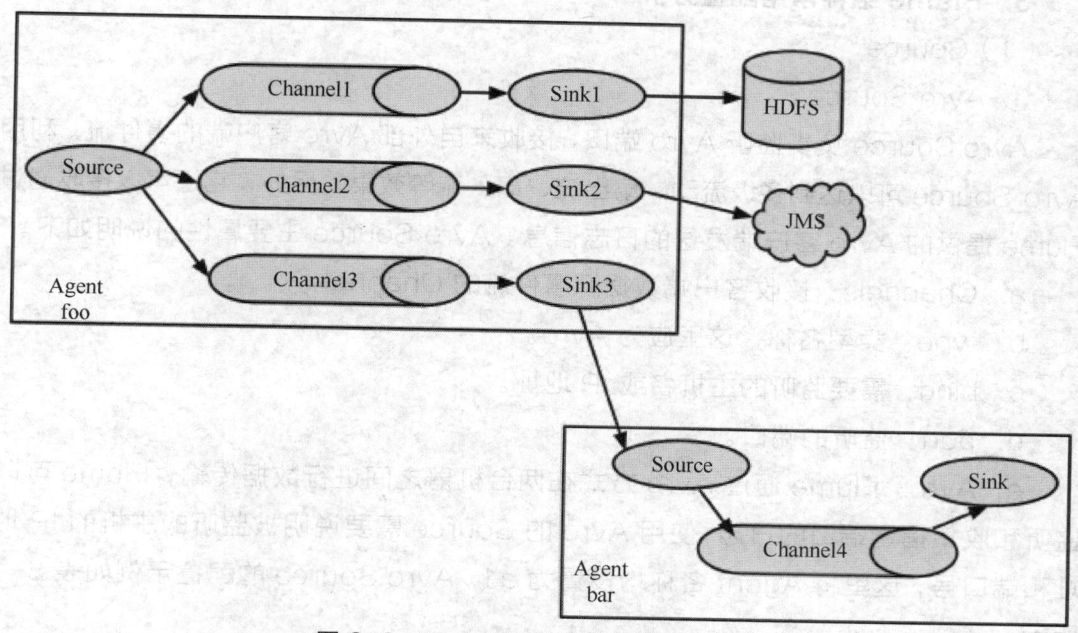

图 3-6 Agent 的常用组合形式（2）

（3）Flume 的可靠性

当 Flume 节点出现故障时，日志能够被传送到其他节点上而不会丢失。Flume 提供了 3 种级别的可靠性保障，从强到弱依次如下。

① end-to-end：收到数据后，Agent 先将 Event 写到磁盘上，当数据传送成功后，再将其删除；如果数据发送失败，则可以重新发送。

② Store on failure：当数据接收方崩溃时，将数据写到本地，待恢复后，继续发送。

③ Best effort：数据发送到接收方后，不会进行确认。

（4）Flume 插件

① Interceptors（拦截器）：用于 Source 和 Channel 之间，用来更改或者检查 Flume 的 Event 数据。

② Channels Selectors（管道选择器）：用于选择使用哪一条管道来传递数据。管道选择器又分为如下两种。

a. 默认管道选择器：每一个管道传递的都是相同的 Event。

b. 多路复用管道选择器：根据每一个 Event 的头部地址选择管道。

③ Sink：用于激活被选择的 Sinks 群中特定的 Sink，实现负载均衡控制。

3. Flume 组件常用配置分析

（1）Source

① Avro Source。

Avro Source 负责监听 Avro 端口，接收来自外部 Avro 客户端的事件流。利用 Avro Source 可以达到多级流动、扇出流、扇入流等效果。另外，它还可以接收通过 Flume 提供的 Avro 客户端发送的日志信息。Avro Source 主要属性的说明如下。

a．Channels：接收客户端数据源事件流的 Channel。

b．type：类型名称，这里应为 Avro。

c．bind：需要监听的主机名或 IP 地址。

d．port：监听的端口。

e．Avro：Flume 通过 Avro 方式在两台机器之间进行数据传输，Flume 可以监听和收集指定端口的日志，使用 Avro 的 Source 需要说明被监听的主机的 IP 地址和端口号，这里将 Agent 名称均设置为 a1。Avro Source 的配置示例如表 3-1 所示。

表 3-1 Avro Source 的配置示例

配置示例	示例说明
a1.Sources = r1	指定采集数据源的名称为 r1
a1.channels = c1	指定使用 Channel 的名称为 c1
a1.Sources.r1.type = Avro	指定采集数据源的方式为 Avro
a1.Sources.r1.channels = c1	指定采集数据源所使用的 Channel
a1.Sources.r1.bind = 0.0.0.0	指定 Avro 监听的主机名或 IP 地址
a1.Sources.r1.port = 4141	指定 Avro 监听的端口

② Exec Source。

Exec Source 可以通过指定的操作对日志进行读取，使用 Exec 时需要指定 shell 命令。Exec Source 的配置示例如表 3-2 所示。

表 3-2 Exec Source 的配置示例

配置示例	示例说明
a1.Sources = r1	指定采集数据源的名称为 r1
a1.channels = c1	指定使用 Channel 的名称为 c1
a1.Sources.r1.type = Exec	指定采集数据源的方式为 Exec

续表

配置示例	示例说明
a1.Sources.r1.command=tail-F var/log/secure	指定 Exec 执行的 shell 命令
a1.Sources.r1.channels = c1	指定采集数据源所使用的 Channel 为 c1

③ Spooling-directory Source。

此 Source 允许将要收集的数据放置到指定的“搜集目录”中。它会监视该目录，并解析出现的新文件。此 Source 的事件处理逻辑是可插拔的，当一个文件被完全读入 Channel 时，它会被重命名或直接删除。

需要注意的是，放置到“搜集目录”中的文件无法修改，如果修改，则 Flume 会报错。另外，也不能产生重名的文件，如果有重名的文件被放置进来，则 Flume 也会报错。Spooling-directory Source 主要属性的说明如下。

Channels：接收客户端数据源事件流的 Channel。

type：类型名称，这里为 spooldir。

spooldir：读取文件的路径，即指定的“搜集目录”。

COMPLETED：对处理完成的文件追加的后缀。

Flume 可以读取 spooldir 对应文件夹中的日志，使用时会指定一个文件夹映射到 spooldir，Flume 可以读取该文件夹中的所有文件。Spooling-directory Source 的配置示例如表 3-3 所示。

表 3-3　Spooling-directory Source 的配置示例

配置示例	示例说明
a1.Sources = r1	指定采集数据源的名称为 r1
a1.channels = c1	指定使用 Channel 的名称为 c1
a1.Sources.r1.type = spooldir	指定采集数据源的类型为 spooldir
a1.Sources.r1.spoolDir = /var/log/apache/flumeSpool	指定 spoolDir 所监视的文件夹
a1.Sources.r1.channels = c1	指定采集数据源所使用的 Channel 为 c1
a1.Sources.r1.fileHeader = true	指定采集过程中将数据源保存成文件以提高容错

④ Syslog Source。

Syslog Source 可以通过 Syslog 协议读取系统日志，采集数据源的类型为 Syslogudp 和 Syslogtcp 两种，使用时需指定 IP 地址和端口。Syslog Source 的配置示例如表 3-4 所示。

表 3-4 Syslog Source 的配置示例

配置示例	示例说明
a1.Sources = r1	指定采集数据源的名称为 r1
a1.channels = c1	指定使用 Channel 的名称为 c1
a1.Sources.r1.type = Syslogudp	指定采集数据源的类型为 Syslogudp
a1.Sources.r1.host = localhost	指定 Syslog 所监视的主机名为 localhost
a1.Sources.r1.port = 5140	指定 Syslog 所监视的端口为 5140
a1.Sources.r1.channels = c1	指定采集数据源所使用的 Channel 为 c1

（2）Channel

Flume 的 Channel 种类并不多，最常用的是 memory channel。其配置示例如表 3-5 所示。

表 3-5 memory channel 的配置示例

配置示例	示例说明
a1.channels = c1	指定使用 Channel 的名称为 c1
a1.channels.c1.type = memory	指定 Channel 的类型为 memory
a1.channels.c1.capacity = 10000	指定 Channel 的容量为 10000
a1.channels.c1.transactionCapacity = 10000	指定 Channel 提交数据的阈值
a1.channels.c1.byteCapacityBufferPercentage = 20	定义 Channel 中 Event 所占的百分比，需要考虑在 Header 中的数据
a1.channels.c1.byteCapacity = 800000	指定 Channel c1 的 byteCapacity（字节数据阈值）为 800000

（3）Sink

① Logger Sink。

Logger Sink 用于记录 INFO 级别日志的汇聚点，即将收集到的日志写到 Flume 的 Log 中，是十分简单但非常实用的 Sink，一般用于调试，前面介绍 Source 时用到的 Sink 都是 Logger Sink。Logger Sink 的配置示例如表 3-6 所示。

表 3-6 Logger Sink 的配置示例

配置示例	示例说明
a1.Sinks = k1	指定使用 Sink 的名称为 k1
a1.Sinks.k1.type = Logger	指定使用 Sink 的类型为 Logger

② Avro Sink。

Avro Sink 可以将接收到的日志发送到指定端口，以供级联 Agent 的下一跳收集和接收日志，使用时需要指定目的 IP 地址和端口。Avro Sink 的配置示例如表 3-7 所示。

表 3-7 Avro Sink 的配置示例

配置示例	示例说明
a1.channels = c1	指定使用 Channel 的名称为 c1
a1.Sinks = k1	指定使用 Sink 的名称为 k1
a1.Sinks.k1.type = Avro	指定 Sink 的类型为 Avro
a1.Sinks.k1.channel = c1	指定 Sink 使用的 Channel 为 c1
a1.Sinks.k1.hostname = 127.0.0.1	指定 Avro 的输出主机 IP 地址为 127.0.0.1
a1.Sinks.k1.port = 4545	指定 Avro 的输出端口为 4545

③ File_roll Sink。

File_roll Sink 可以将一定时间内收集到的日志写到一个指定的文件中，具体过程为用户指定一个文件夹和一个周期，再启动 Agent，此时，该文件夹会产生一个文件，并将该周期内收集到的日志全部写入到该文件中，直到下一个周期再次产生一个新文件并继续写入，以此类推，周而复始。File_roll Sink 的配置示例如表 3-8 所示。

表 3-8 File_roll Sink 的配置示例

配置示例	示例说明
a1.channels = c1	指定使用 Channel 的名称为 c1
a1.Sinks = k1	指定使用 Sink 的名称为 k1
a1.Sinks.k1.type = File_roll	指定 Sink 的类型为 File_roll

任务实施

1. Flume 的安装

打开已经安装好的 Ubuntu 系统（JDK 需自行解压安装），执行“cd”命令，进入保存 Flume 安装包的大数据根目录（这里的大数据根目录为/data/bigdata），并执行“ls”命令，查看 Flume 安装包，如图 3-7 所示。

```
hadoop@master:/usr/local/hadoop/sbin$ cd /data/bigdata/
hadoop@master:/data/bigdata$ ls
apache-flume-1.8.0-bin.tar.gz
```

图 3-7 查看 Flume 安装包

执行“tar”命令，解压 Flume 安装包 apache-flume-1.8.0-bin.tar.gz，如图 3-8 所示。

```
hadoop@master:/data/bigdata$ sudo tar zxvf apache-flume-1.8.0-bin.tar.gz
apache-flume-1.8.0-bin/lib/flume-ng-configuration-1.8.0.jar
apache-flume-1.8.0-bin/lib/slf4j-api-1.6.1.jar
apache-flume-1.8.0-bin/lib/slf4j-log4j12-1.6.1.jar
apache-flume-1.8.0-bin/lib/log4j-1.2.17.jar
apache-flume-1.8.0-bin/lib/guava-11.0.2.jar
apache-flume-1.8.0-bin/lib/jsr305-1.3.9.jar
apache-flume-1.8.0-bin/lib/flume-ng-sdk-1.8.0.jar
apache-flume-1.8.0-bin/lib/avro-1.7.4.jar
apache-flume-1.8.0-bin/lib/jackson-core-asl-1.9.3.jar
apache-flume-1.8.0-bin/lib/jackson-mapper-asl-1.9.3.jar
apache-flume-1.8.0-bin/lib/paranamer-2.3.jar
apache-flume-1.8.0-bin/lib/snappy-java-1.1.4.jar
apache-flume-1.8.0-bin/lib/commons-compress-1.4.1.jar
apache-flume-1.8.0-bin/lib/xz-1.0.jar
apache-flume-1.8.0-bin/lib/avro-ipc-1.7.4.jar
apache-flume-1.8.0-bin/lib/jetty-6.1.26.jar
apache-flume-1.8.0-bin/lib/jetty-util-6.1.26.jar
apache-flume-1.8.0-bin/lib/velocity-1.7.jar
apache-flume-1.8.0-bin/lib/commons-collections-3.2.2.jar
apache-flume-1.8.0-bin/lib/commons-lang-2.5.jar
apache-flume-1.8.0-bin/lib/netty-3.9.4.Final.jar
apache-flume-1.8.0-bin/lib/libthrift-0.9.3.jar
```

图 3-8 解压 Flume 安装包

进入 Flume 根目录，查看 Flume 中的文件，如图 3-9 所示。

```
hadoop@master:/data/bigdata$ cd apache-flume-1.8.0-bin/
hadoop@master:/data/bigdata/apache-flume-1.8.0-bin$ ll
total 160
drwxr-xr-x  7 root root   4096 Aug  9 15:20 ./
drwxrwxrwx  3 root root   4096 Aug  9 15:20 ../
drwxr-xr-x  2 root staff  4096 Aug  9 15:20 bin/
-rw-r--r--  1 root staff 81264 Sep 15  2017 CHANGELOG
drwxr-xr-x  2 root staff  4096 Aug  9 15:20 conf/
-rw-r--r--  1 root staff  5681 Sep 15  2017 DEVNOTES
-rw-r--r--  1 root staff  2873 Sep 15  2017 doap_Flume.rdf
drwxr-xr-x 10 root root   4096 Sep 15  2017 docs/
drwxr-xr-x  2 root root   4096 Aug  9 15:20 lib/
-rw-r--r--  1 root staff 27663 Sep 15  2017 LICENSE
-rw-r--r--  1 root staff   249 Sep 15  2017 NOTICE
-rw-r--r--  1 root staff  2483 Sep 15  2017 README.md
-rw-r--r--  1 root staff  1588 Sep 15  2017 RELEASE-NOTES
drwxr-xr-x  2 root root   4096 Aug  9 15:20 tools/
```

图 3-9 查看 Flume 中的文件

bin 目录中存放的是 Flume 的可执行文件，conf 目录中存放的是 Flume 的配置文件，lib 目录中存放的是 Flume 的依赖 JAR 包。

为了方便操作路径引用，需要在/etc/profile 中配置相关环境变量，代码如下。

```
# 配置 JDK 环境变量
export JAVA_HOME=/usr/local/jdk1.8
export CLASSPATH=$:CLASSPATH:JAVA_HOME/lib/
```

```
export PATH=$PATH:$JAVA_HOME/bin
# 配置 Hadoop 环境变量
export HADOOP_HOME=/usr/local/Hadoop
export path=$PATH:$HADOOP_HOME/bin:$HADOOP_HOME/sbin
# 配置 Flume 环境变量
export FLUME_HOME=/data/bigdata/apache-flume-1.8.0-bin
export PATH=.:$PATH:$FLUME_HOME/bin
```

进入 conf 目录，通过执行“cp”命令复制 conf 目录中的 flume-env.sh.template 文件，创建的副本文件名称为 flume-env.sh，如图 3-10 所示。

```
hadoop@master:~$ cd /data/bigdata/
hadoop@master:/data/bigdata$ cd apache-flume-1.8.0-bin/
hadoop@master:/data/bigdata/apache-flume-1.8.0-bin$ cd conf/
hadoop@master:/data/bigdata/apache-flume-1.8.0-bin/conf$ ll
total 24
drwxr-xr-x 2 hadoop staff 4096 Aug  9 15:20 ./
drwxr-xr-x 7 hadoop root  4096 Aug  9 15:20 ../
-rw-r--r-- 1 hadoop staff 1661 Sep 15  2017 flume-conf.properties.template
-rw-r--r-- 1 hadoop staff 1455 Sep 15  2017 flume-env.ps1.template
-rw-r--r-- 1 hadoop staff 1568 Sep 15  2017 flume-env.sh.template
-rw-r--r-- 1 hadoop staff 3107 Sep 15  2017 log4j.properties
hadoop@master:/data/bigdata/apache-flume-1.8.0-bin/conf$ sudo cp flume-env.sh.te
mplate flume-env.sh
[sudo] password for hadoop:
hadoop@master:/data/bigdata/apache-flume-1.8.0-bin/conf$ ll
total 28
drwxr-xr-x 2 hadoop staff 4096 Aug  9 15:26 ./
drwxr-xr-x 7 hadoop root  4096 Aug  9 15:20 ../
-rw-r--r-- 1 hadoop staff 1661 Sep 15  2017 flume-conf.properties.template
-rw-r--r-- 1 hadoop staff 1455 Sep 15  2017 flume-env.ps1.template
-rw-r--r-- 1 root   root  1568 Aug  9 15:26 flume-env.sh
-rw-r--r-- 1 hadoop staff 1568 Sep 15  2017 flume-env.sh.template
-rw-r--r-- 1 hadoop staff 3107 Sep 15  2017 log4j.properties
```

图 3-10　复制文件并修改其名称

编辑配置文件 flume-env.sh，修改 JAVA_HOME 的值为“/usr/local/jdk1.8”，如图 3-11 所示。

执行“flume-ng version”命令，验证 Flume 是否安装成功，如果进入图 3-12 所示界面，表示 Flume 安装成功。

2. 企业常用 Flume 配置

（1）互联网公司采集日志时，往往需要将各个 Web 服务器的日志汇总到一台日志分析服务器中，以便日志分析。可通过 Flume 监听指定自身端口来采集其他设备客户端发送到指定端口的数据，即使用 Flume 将 slave1 节点的文件采集到 master 的指定目录中。

```
#
# Unless required by applicable law or agreed to in writing, software
# distributed under the License is distributed on an "AS IS" BASIS,
# WITHOUT WARRANTIES OR CONDITIONS OF ANY KIND, either express or implied.
# See the License for the specific language governing permissions and
# limitations under the License.

# If this file is placed at FLUME_CONF_DIR/flume-env.sh, it will be sourced
# during Flume startup.

# Enviroment variables can be set here.

export JAVA_HOME=/usr/local/jdk1.8

# Give Flume more memory and pre-allocate, enable remote monitoring via JMX
# export JAVA_OPTS="-Xms100m -Xmx2000m -Dcom.sun.management.jmxremote"

# Let Flume write raw event data and configuration information to its log files
for debugging
# purposes. Enabling these flags is not recommended in production,
# as it may result in logging sensitive user information or encryption secrets.
# export JAVA_OPTS="$JAVA_OPTS -Dorg.apache.flume.log.rawdata=true -Dorg.apache.
flume.log.printconfig=true "
```

图 3-11　编辑配置文件

```
hadoop@master:/data/bigdata/apache-flume-1.8.0-bin/bin$ ./flume-ng version
Flume 1.8.0
Source code repository: https://git-wip-us.apache.org/repos/asf/flume.git
Revision: 99f591994468633fc6f8701c5fc53e0214b6da4f
Compiled by denes on Fri Sep 15 14:58:00 CEST 2017
From source with checksum fbb44c8c8fb63a49be0a59e27316833d
```

图 3-12　Flume 安装成功

在 master 节点下进行如下操作。

进入 Flume 的 conf 目录，创建 Avro.conf 配置文件，如图 3-13 所示。

```
hadoop@master:~$ cd /data/bigdata/apache-flume-1.8.0-bin/conf
hadoop@master:/data/bigdata/apache-flume-1.8.0-bin/conf$ sudo vim Avro.conf
```

图 3-13　创建 Avro.conf 配置文件

编辑 Avro.conf 配置文件，如图 3-14 所示。

```
# Avro.conf 配置文件
# 命名 Agent a1 的组件
a1.Sources = r1
a1.Sinks = k1
a1.channels = c1
# 配置 Source 信息
a1.Sources.r1.type = Avro
a1.Sources.r1.channels = c1
a1.Sources.r1.bind = 0.0.0.0
a1.Sources.r1.port = 4141
# Sink 信息
a1.Sinks.k1.type = logger
# 配置 Channels 的类型为 memory
a1.channels.c1.type = memory
a1.channels.c1.capacity = 1000
a1.channels.c1.transactionCapacity = 100
# 将 Source、Sink 绑定到 Channel
a1.Sources.r1.channels = c1
a1.Sinks.k1.channel = c1
```

图 3-14　编辑 Avro.conf 配置文件

通过执行命令“flume-ng agent -c . -f /data/bigdata/apache-flume-1.8.0-bin/conf/avro.conf -n a1 -DFlume.root.logger=INFO,console”启动 Flume Agent a1，指定日志等级为 INFO，并将日志输出到控制台，如图 3-15 所示。

```
hadoop@master:/data/bigdata/apache-flume-1.8.0-bin/bin$ ./flume-ng agent -c . -f
 /data/bigdata/apache-flume-1.8.0-bin/conf/avro.conf -n a1 -DFlume.root.logger=I
NFO,console
Info: Including Hadoop libraries found via (/usr/local/hadoop/bin/hadoop) for HD
FS access
Info: Including Hive libraries found via () for Hive access
+ exec /usr/local/jdk1.8/bin/java -Xmx20m -DFlume.root.logger=INFO,console -cp '
/data/bigdata/apache-flume-1.8.0-bin/bin:/data/bigdata/apache-flume-1.8.0-bin/li
b/*:/usr/local/hadoop/etc/hadoop:/usr/local/hadoop/share/hadoop/common/lib/*:/us
r/local/hadoop/share/hadoop/common/*:/usr/local/hadoop/share/hadoop/hdfs:/usr/lo
cal/hadoop/share/hadoop/hdfs/lib/*:/usr/local/hadoop/share/hadoop/hdfs/*:/usr/lo
cal/hadoop/share/hadoop/yarn/lib/*:/usr/local/hadoop/share/hadoop/yarn/*:/usr/lo
cal/hadoop/share/hadoop/mapreduce/lib/*:/usr/local/hadoop/share/hadoop/mapreduce
/*:/usr/local/hadoop/contrib/capacity-scheduler/*.jar:/lib/*' -Djava.library.pat
h=:/usr/local/hadoop/lib/native org.apache.flume.node.Application -f /data/bigda
ta/apache-flume-1.8.0-bin/conf/avro.conf -n a1
SLF4J: Class path contains multiple SLF4J bindings.
SLF4J: Found binding in [jar:file:/data/bigdata/apache-flume-1.8.0-bin/lib/slf4j
-log4j12-1.6.1.jar!/org/slf4j/impl/StaticLoggerBinder.class]
SLF4J: Found binding in [jar:file:/usr/local/hadoop/share/hadoop/common/lib/slf4
j-log4j12-1.7.10.jar!/org/slf4j/impl/StaticLoggerBinder.class]
```

图 3-15　启动 Flume Agent a1

打开一个新的控制台命令行窗口，在 slave1 节点下进行如下操作。

在 Flume 目录中创建一个自定义名称的文件夹，并创建一个要发送的测试文件（由主机端采集），如图 3-16 所示。

```
hadoop@slave1:/data/bigdata/apache-flume-1.8.0-bin$ sudo mkdir 20yyxxxxABC
hadoop@slave1:/data/bigdata/apache-flume-1.8.0-bin$ cd 20yyxxxxABC/
hadoop@slave1:/data/bigdata/apache-flume-1.8.0-bin/20yyxxxxABC$ sudo touch log.00
hadoop@slave1:/data/bigdata/apache-flume-1.8.0-bin/20yyxxxxABC$ sudo echo "hello world" > log.00

hadoop@slave1:/data/bigdata/apache-flume-1.8.0-bin/20yyxxxxABC$ ll
total 12
drwxr-xr-x 2 root   root 4096 Aug  9 16:13 ./
drwxr-xr-x 8 root   root 4096 Aug  9 16:13 ../
-rw-r--r-- 1 hadoop root   12 Aug  9 16:14 log.00
```

图 3-16　创建测试文件

使用 echo 向文件中输出内容的基本方法是使用 IO 重定向指令“>”，默认情况下 echo 输出到标准输出中，使用“>”指令时可重定向输出到文件中。

启动 Flume Avro-client，并指定主机的接收端口为 4141，向主机 master 发送日志信息数据，如图 3-17 所示。

```
Flume-ng Avro-client -c . -H master -p 4141 -F /data/bigdata/apache-Flume-1.8.0-bin/20
yyxxxxABC/log.00
```

图 3-17　向主机 master 发送日志信息数据

查看 master 节点在命令行窗口中接收的数据，如图 3-18 所示。

```
18/08/09 15:39:31 INFO source.AvroSource: Starting Avro source r1: { bindAddress
: 192.168.1.14, port: 4141 }...
18/08/09 15:39:31 INFO instrumentation.MonitoredCounterGroup: Monitored counter
group for type: SOURCE, name: r1: Successfully registered new MBean.
18/08/09 15:39:31 INFO instrumentation.MonitoredCounterGroup: Component type: SO
URCE, name: r1 started
18/08/09 15:39:31 INFO source.AvroSource: Avro source r1 started.
18/08/09 16:17:34 INFO ipc.NettyServer: [id: 0x139e12e9, /192.168.1.93:33658 =>
/192.168.1.14:4141] OPEN
18/08/09 16:17:34 INFO ipc.NettyServer: [id: 0x139e12e9, /192.168.1.93:33658 =>
/192.168.1.14:4141] BOUND: /192.168.1.14:4141
18/08/09 16:17:34 INFO ipc.NettyServer: [id: 0x139e12e9, /192.168.1.93:33658 =>
/192.168.1.14:4141] CONNECTED: /192.168.1.93:33658
18/08/09 16:17:35 INFO ipc.NettyServer: [id: 0x139e12e9, /192.168.1.93:33658 :>
/192.168.1.14:4141] DISCONNECTED
18/08/09 16:17:35 INFO ipc.NettyServer: [id: 0x139e12e9, /192.168.1.93:33658 :>
/192.168.1.14:4141] UNBOUND
18/08/09 16:17:35 INFO ipc.NettyServer: [id: 0x139e12e9, /192.168.1.93:33658 :>
/192.168.1.14:4141] CLOSED
18/08/09 16:17:35 INFO ipc.NettyServer: Connection to /192.168.1.93:33658 discon
nected.
18/08/09 16:17:38 INFO sink.LoggerSink: Event: { headers:{} body: 68 65 6C 6C 6F
 20 77 6F 72 6C 64                hello world }
```

图 3-18　查看 master 节点在命令行窗口中接收的数据

观察到日志文件已经通过 Flume 发送到了 master 节点。按快捷键“Ctrl+C”退出。

以上还原了企业中使用 Flume Avro 进行数据传输的过程，通过 Avro-client 发送数据到接收端的指定端口，实现了日志的传递。在企业信息采集中可以将分散在各个机器上的日志汇总到日志处理分析服务器中，从而完成整体数据的采集。

（2）为了自动采集每日产生的日志，互联网公司一般采用日志策略：每日自动生成一个日志文件。为了实现自动采集日志功能，Flume 提供了 Spooling Directory Source 方式来对日志存储目录进行监控，收集新生成的文件，即使用 Flume 监控本地目录，如有新增文件就采集文件中的数据。

在 Flume 目录中创建一个自定义名称的文件夹，在该文件夹中创建一个被监听的文件夹 spool，如图 3-19 所示。

```
hadoop@master:/data/bigdata/apache-flume-1.8.0-bin$ sudo mkdir 20yyxxxxABC
[sudo] password for hadoop:
hadoop@master:/data/bigdata/apache-flume-1.8.0-bin$ cd 20yyxxxxABC/
hadoop@master:/data/bigdata/apache-flume-1.8.0-bin/20yyxxxxABC$ sudo mkdir spool
hadoop@master:/data/bigdata/apache-flume-1.8.0-bin/20yyxxxxABC$ ll
total 12
drwxr-xr-x 3 root   root 4096 Aug  9 16:19 ./
drwxr-xr-x 8 hadoop root 4096 Aug  9 16:19 ../
drwxr-xr-x 2 root   root 4096 Aug  9 16:19 spool/
```

图 3-19　创建被监听的文件夹 spool

进入 Flume 的 conf 目录，创建 spool.conf 配置文件，如图 3-20 所示。

```
hadoop@master:/data/bigdata/apache-flume-1.8.0-bin/conf$ sudo vim spool.conf
```

图 3-20　创建 spool.conf 配置文件

编辑 spool.conf 配置文件，如图 3-21 所示。

```
# spool.conf 配置文件
a2.Sources = r2
a2.Sinks = k2
a2.channels = c2
# 配置 Source 信息
a2.Sources.r2.type = spooldir
a2.Sources.r2.channels = c2
a2.Sources.r2.spoolDir = /data/bigdata/apache-flume-1.8.0-bin/20yyxxxxABC/spool
a2.Sources.r2.fileHeader = true
# 配置 Sink 信息
a2.Sinks.k2.type = logger
# 配置 Channel 信息
a2.channels.c2.type = memory
a2.channels.c2.capacity = 1000
a2.channels.c2.transactionCapacity = 100
```

图 3-21　编辑 spool.conf 配置文件

通过执行命令“flume-ng agent -c . -f /data/bigdata/apache-Flume-1.8.0-bin/conf/spool.conf -n a2 -DFlume.root.logger=INFO,console”启动 Flume Agent a2，指定日志等级为 INFO，并将日志内容输出到控制台，Flume Agent a2 的启动信息如图 3-22 所示。

```
18/08/09 16:30:33 INFO channel.DefaultChannelFactory: Creating instance of chann
el c2 type memory
18/08/09 16:30:33 INFO node.AbstractConfigurationProvider: Created channel c2
18/08/09 16:30:33 INFO source.DefaultSourceFactory: Creating instance of source
r2, type spooldir
18/08/09 16:30:33 INFO node.AbstractConfigurationProvider: Channel c2 connected
to [r2]
18/08/09 16:30:33 INFO node.Application: Starting new configuration:{ sourceRunn
ers:{r2=EventDrivenSourceRunner: { source:Spool Directory source r2: { spoolDir:
 /data/bigdata/apache-flume-1.8.0-bin/20yyxxxxABC/spool } }} sinkRunners:{} chan
nels:{c2=org.apache.flume.channel.MemoryChannel{name: c2}} }
18/08/09 16:30:33 INFO node.Application: Starting Channel c2
18/08/09 16:30:33 INFO instrumentation.MonitoredCounterGroup: Monitored counter
group for type: CHANNEL, name: c2: Successfully registered new MBean.
18/08/09 16:30:33 INFO instrumentation.MonitoredCounterGroup: Component type: CH
ANNEL, name: c2 started
18/08/09 16:30:33 INFO node.Application: Starting Source r2
18/08/09 16:30:33 INFO source.SpoolDirectorySource: SpoolDirectorySource source
starting with directory: /data/bigdata/apache-flume-1.8.0-bin/20yyxxxxABC/spool
18/08/09 16:30:33 INFO instrumentation.MonitoredCounterGroup: Monitored counter
group for type: SOURCE, name: r2: Successfully registered new MBean.
18/08/09 16:30:33 INFO instrumentation.MonitoredCounterGroup: Component type: SO
URCE, name: r2 started
```

图 3-22　Flume Agent a2 的启动信息

打开一个新的命令行窗口，用来观察日志。在被监听的文件目录中创建一个包含内容的测试文件，如图 3-23 所示。

```
hadoop@master:~$ sudo echo "hello world2" > /data/bigdata/apache-flume-1.8.0-bin
/20yyxxxxABC/spool/log.01
```

图 3-23　创建包含内容的测试文件

随后可以观察第一个命令行窗口的变化，查看接收的数据，如图 3-24 所示。

```
18/08/09 16:33:25 INFO avro.ReliableSpoolingFileEventReader: Last read took us j
ust up to a file boundary. Rolling to the next file, if there is one.
18/08/09 16:33:25 INFO avro.ReliableSpoolingFileEventReader: Preparing to move f
ile /data/bigdata/apache-flume-1.8.0-bin/20yyxxxxABC/spool/log.01 to /data/bigda
ta/apache-flume-1.8.0-bin/20yyxxxxABC/spool/log.01.COMPLETED
```

图 3-24 查看接收的数据

观察到日志文件已经被监听到了，按快捷键“Ctrl+C”退出。

需要注意，复制到 spool 目录中的文件不可以再打开进行编辑，spool 目录中不可包含相应的子目录。此外，还要注意路径的长度，路径过长有可能导致 Flume 无法正常运行。

以上还原了企业中使用 Flume Spooling Directory 来监视目录中每天新日志生成的场景，以及对新产生的日志实现自动采集的过程。

（3）互联网公司通常使用前两种方式就可以基本完成日志数据的采集和汇总，但在实际工作中，对于不断变化的实时生成的日志，仅仅依靠 Avro 和 Spooling Directory 采集方式，无法实现对日志的实时采集和高可靠采集，因此 Flume 提供了一种 Exec 的方式来自定义日志采集，即通过使用 Exec Source 完成从本地日志文件中收集日志数据的任务。

进入 Flume 的“自行创建”目录，创建一个空的数据源文件 log.02。

进入 Flume 的 conf 目录，创建 exec_tail.conf 配置文件，如图 3-25 所示。

```
hadoop@master:/data/bigdata/apache-flume-1.8.0-bin/conf$ sudo vim exec_tail.conf
```

图 3-25 创建 exec_tail.conf 配置文件

编辑 exec_tail.conf 配置文件，如图 3-26 所示。

```
a3.Sources = r3
a3.Sinks = k3
a3.channels = c3
# 配置 Source 信息
a3.Sources.r3.type = exec
a3.Sources.r3.channels = c3
# 注意，下面代码中包含要监控的日志文件的路径
a3.Sources.r3.command = tail -F /data/bigdata/apache-flume-1.8.0-bin/20yyxxxxABC/log.02
# 配置 Sink 信息
a3.Sinks.k3.type = logger
# 使用缓冲内存中事件的 Channel
a3.channels.c3.type = memory
a3.channels.c3.capacity = 1000
a3.channels.c3.transactionCapacity = 100
# 将 Source，Sink 绑定到 Channel 中
a3.Sources.r3.channels = c3
a3.Sinks.k3.channel = c3
```

图 3-26 编辑 exec_tail.conf 配置文件

通过执行命令“flume-ng agent -c . -f /data/bigdata/apache-flume-1.8.0-bin/conf/exec_tail.conf -n a3- DFlume.root.logger=INFO,console”启动 Flume Agent a3，指定日志等级为 INFO，并将日志信息输出到控制台。Flume Agent a3 的启动信息如图 3-27 所示。

```
18/08/09 17:05:22 INFO sink.DefaultSinkFactory: Creating instance of sink: k3, t
ype: logger
18/08/09 17:05:22 INFO node.AbstractConfigurationProvider: Channel c3 connected
to [r3, k3]
18/08/09 17:05:22 INFO node.Application: Starting new configuration:{ sourceRunn
ers:{r3=EventDrivenSourceRunner: { source:org.apache.flume.source.ExecSource{nam
e:r3,state:IDLE} }} sinkRunners:{k3=SinkRunner: { policy:org.apache.flume.sink.D
efaultSinkProcessor@7012b8f0 counterGroup:{ name:null counters:{} } }} channels:
{c3=org.apache.flume.channel.MemoryChannel{name: c3}} }
18/08/09 17:05:22 INFO node.Application: Starting Channel c3
18/08/09 17:05:22 INFO instrumentation.MonitoredCounterGroup: Monitored counter
group for type: CHANNEL, name: c3: Successfully registered new MBean.
18/08/09 17:05:22 INFO instrumentation.MonitoredCounterGroup: Component type: CH
ANNEL, name: c3 started
18/08/09 17:05:22 INFO node.Application: Starting Sink k3
18/08/09 17:05:22 INFO node.Application: Starting Source r3
18/08/09 17:05:22 INFO source.ExecSource: Exec source starting with command: tai
l -F /usr/local/hadoop/log_exec_tail
18/08/09 17:05:22 INFO instrumentation.MonitoredCounterGroup: Monitored counter
group for type: SOURCE, name: r3: Successfully registered new MBean.
18/08/09 17:05:22 INFO instrumentation.MonitoredCounterGroup: Component type: SO
URCE, name: r3 started
```

图 3-27　Flume Agent a3 的启动信息

执行“for i in {1..100};do echo "exec tail$i">> /data/bigdata/apache-flume-1.8.0-bin/20yyxxxxABC/log.02 c;echo $i;sleep 0.01;done”命令，为 log.02 文件循环追加大量数据，观察控制台输出端，发现追加的日志已被获取且输出到屏幕上，日志输出信息如图 3-28 所示。

```
 74 61 69 6C 38 39 20 63          exec tail89 c }
18/08/09 17:15:32 INFO sink.LoggerSink: Event: { headers:{} body: 65 78 65 63 20
 74 61 69 6C 39 30 20 63          exec tail90 c }
18/08/09 17:15:32 INFO sink.LoggerSink: Event: { headers:{} body: 65 78 65 63 20
 74 61 69 6C 39 31 20 63          exec tail91 c }
18/08/09 17:15:32 INFO sink.LoggerSink: Event: { headers:{} body: 65 78 65 63 20
 74 61 69 6C 39 32 20 63          exec tail92 c }
18/08/09 17:15:32 INFO sink.LoggerSink: Event: { headers:{} body: 65 78 65 63 20
 74 61 69 6C 39 33 20 63          exec tail93 c }
18/08/09 17:15:32 INFO sink.LoggerSink: Event: { headers:{} body: 65 78 65 63 20
 74 61 69 6C 39 34 20 63          exec tail94 c }
18/08/09 17:15:32 INFO sink.LoggerSink: Event: { headers:{} body: 65 78 65 63 20
 74 61 69 6C 39 35 20 63          exec tail95 c }
18/08/09 17:15:32 INFO sink.LoggerSink: Event: { headers:{} body: 65 78 65 63 20
 74 61 69 6C 39 36 20 63          exec tail96 c }
18/08/09 17:15:32 INFO sink.LoggerSink: Event: { headers:{} body: 65 78 65 63 20
 74 61 69 6C 39 37 20 63          exec tail97 c }
18/08/09 17:15:32 INFO sink.LoggerSink: Event: { headers:{} body: 65 78 65 63 20
 74 61 69 6C 39 38 20 63          exec tail98 c }
18/08/09 17:15:32 INFO sink.LoggerSink: Event: { headers:{} body: 65 78 65 63 20
 74 61 69 6C 39 39 20 63          exec tail99 c }
18/08/09 17:15:32 INFO sink.LoggerSink: Event: { headers:{} body: 65 78 65 63 20
 74 61 69 6C 31 30 30 20 63       exec tail100 c }
```

图 3-28　日志输出信息

查看被监听目录中的文件，发现文件中增加了“.COMPLETED”标识，如图3-29 所示。

```
hadoop@master:/data/bigdata/apache-flume-1.8.0-bin/20yyxxxxABC$ cd spool/
hadoop@master:/data/bigdata/apache-flume-1.8.0-bin/20yyxxxxABC/spool$ ls
log.01.COMPLETED
```

图 3-29 被监听目录中的文件增加了“.COMPLETED”标识

（4）Flume 提供了 Syslog 的方式，通过 TCP/UDP 通信协议直接对某台主机上的某个端口进行监听，实现了采集端主动采集端口日志的功能，提高了可靠性。

Syslog 可以通过 Syslog 协议读取系统日志，协议分为 TCP 和 UDP 两种，使用时需指定 IP 地址和端口，此任务中使用 Flume 监听本机的 5140 端口。

进入 Flume 的 conf 目录，创建名称为 syslog_tcp.conf 的配置文件，如图 3-30 所示。

```
hadoop@master:/data/bigdata/apache-flume-1.8.0-bin/conf$ sudo vim syslog_tcp.conf
```

图 3-30 创建 syslog_tcp.conf 配置文件

编辑 syslog_tcp.conf 配置文件，如图 3-31 所示。

```
# syslog_tcp.conf 配置文件
a4.Sources = r4
a4.Sinks = k4
a4.channels = c4
# 配置 Source 信息源
a4.Sources.r4.type = Syslogtcp
a4.Sources.r4.port = 5140
a4.Sources.r4.host = localhost
a4.Sources.r4.channels = c4
# 配置 Sink 信息
a4.Sinks.k4.type = logger
# 配置 Channel 信息
a4.channels.c4.type = memory
a4.channels.c4.capacity = 1000
a4.channels.c4.transactionCapacity = 100
# 将 Source、Sink 绑定到 Channel 中
a4.Sources.r4.channels = c4
a4.Sinks.k4.channel = c4
```

图 3-31 编辑 syslog_tcp.conf 配置文件

通过执行命令“flume-ng agent -c . -f/data/bigdata/apache-flume-1.8.0 -bin/conf/syslog_tcp.conf-n a4 -DFlume.root.logger=INFO,console”启动 Flume Agent a4，指定日志等级为 INFO，并将日志内容输出到控制台。Flume Agent a4 的启动信息如图 3-32 所示。

```
18/08/09 17:23:31 INFO sink.DefaultSinkFactory: Creating instance of sink: k4, type
: logger
18/08/09 17:23:31 INFO node.AbstractConfigurationProvider: Channel c4 connected to
[r4, k4]
18/08/09 17:23:31 INFO node.Application: Starting new configuration:{ sourceRunners
:{r4=EventDrivenSourceRunner: { source:org.apache.flume.source.SyslogTcpSource{name
:r4,state:IDLE} }} sinkRunners:{k4=SinkRunner: { policy:org.apache.flume.sink.Defau
ltSinkProcessor@660a3b35 counterGroup:{ name:null counters:{} } }} channels:{c4=org
.apache.flume.channel.MemoryChannel{name: c4}} }
18/08/09 17:23:31 INFO node.Application: Starting Channel c4
18/08/09 17:23:31 INFO node.Application: Waiting for channel: c4 to start. Sleeping
 for 500 ms
18/08/09 17:23:31 INFO instrumentation.MonitoredCounterGroup: Monitored counter gro
up for type: CHANNEL, name: c4: Successfully registered new MBean.
18/08/09 17:23:31 INFO instrumentation.MonitoredCounterGroup: Component type: CHANN
EL, name: c4 started
18/08/09 17:23:32 INFO node.Application: Starting Sink k4
18/08/09 17:23:32 INFO node.Application: Starting Source r4
18/08/09 17:23:32 INFO source.SyslogTcpSource: Syslog TCP Source starting...
18/08/09 17:23:32 INFO instrumentation.MonitoredCounterGroup: Monitored counter gro
up for type: SOURCE, name: r4: Successfully registered new MBean.
18/08/09 17:23:32 INFO instrumentation.MonitoredCounterGroup: Component type: SOURC
E, name: r4 started
```

图 3-32　Flume Agent a4 的启动信息

重新打开一个命令行窗口，使用“nc”命令发送“hello syslog”数据到本机的 5140 端口，如图 3-33 所示。nc 表示 NetCat，是一个非常简单的 Unix 工具，可以读、写 TCP 或 UDP 网络连接。它被设计成一个可靠的后端工具，能被其他程序或脚本直接驱动。同时，它也是一个功能丰富的网络调试和开发工具。

```
hadoop@master:~$ sudo echo "hello syslog" | nc localhost 5140
```

图 3-33　发送数据

观察之前的本机控制台命令行窗口，可以看到追加的日志已被获取，如图 3-34 所示。

```
18/08/09 17:23:31 INFO instrumentation.MonitoredCounterGroup: Monitored counter gro
up for type: CHANNEL, name: c4: Successfully registered new MBean.
18/08/09 17:23:31 INFO instrumentation.MonitoredCounterGroup: Component type: CHANN
EL, name: c4 started
18/08/09 17:23:32 INFO node.Application: Starting Sink k4
18/08/09 17:23:32 INFO node.Application: Starting Source r4
18/08/09 17:23:32 INFO source.SyslogTcpSource: Syslog TCP Source starting...
18/08/09 17:23:32 INFO instrumentation.MonitoredCounterGroup: Monitored counter gro
up for type: SOURCE, name: r4: Successfully registered new MBean.
18/08/09 17:23:32 INFO instrumentation.MonitoredCounterGroup: Component type: SOURC
E, name: r4 started
18/08/09 17:24:35 WARN source.SyslogUtils: Event created from Invalid Syslog data.
18/08/09 17:24:35 INFO sink.LoggerSink: Event: { headers:{Severity=0, Facility=0, f
lume.syslog.status=Invalid} body: 68 65 6C 6C 6F 20 73 79 73 6C 6F 67             h
ello syslog }
```

图 3-34　追加的日志已被获取

任务 2　Flume 采集数据上传到集群

任务描述

（1）学习 Flume 日志采集的相关基础知识。

（2）使用 Flume 采集数据并存储到 HDFS 中。

（3）使用 Flume 采集数据并存储到 HBase 中。

任务目标

（1）熟悉 Flume 的相关基础知识。

（2）学会将采集的日志数据转存到 HDFS 中的方法。

（3）学会将采集的日志数据转存到 HBase 中的方法。

知识准备

1. HDFS Sinks

HDFS 与 File_roll 有些类似，都是将收集到的日志写入到新创建的文件中保存起来，其区别如下：File_roll 的文件存储路径为系统的本地路径；而 HDFS 的存储路径为分布式的文件系统 HDFS 的路径，且 HDFS 创建新文件的周期可以是时间，也可以是文件的大小，还可以是采集日志的条数。HDFS Sinks 的配置示例如表 3-9 所示。

表 3-9　HDFS Sinks 的配置示例

配置示例	示例说明
a1.channels = c1	指定使用 Channel 的名称为 c1
a1.Sinks = k1	指定使用 Sink 的名称为 k1
a1.Sinks.k1.type = hdfs	指定 Sink 的类型为 hdfs
a1.Sinks.k1.channel = c1	指定 Sink 使用的 Channel 为 c1
a1.Sinks.k1.hdfs.path = /flume/events/ %y-%m- %d/%H%M/%S	指定存储的 HDFS 路径，可使用%、{}、时间表达式等作为路径
a1.Sinks.k1.hdfs.filePrefix = events-	指定输出文件的前缀
a1.Sinks.k1.hdfs.round = true	启用时间上的“舍弃”，即 hdfs.path 解析时将根据 roundValue 中设置的时间进行舍弃，时间单位根据 roundUnit 中的设置来确定
a1.Sinks.k1.hdfs.roundValue = 10	设定时间值
a1.Sinks.k1.hdfs.roundUnit = minute	设定时间单位

2. HBaseSinks

Flume 有两大类 HBaseSinks：HBaseSink 和 AsyncHBaseSink。

（1）HBaseSink

HBaseSink 提供了两种序列化模式：SimpleHbaseEventSerializer 和 RegexHbaseEventSerializer。SimpleHbaseEventSerializer 将整个事件的 body 部分当作完整的一列写入 HBase，因此在插入 HBase 的时候，一个事件的 body 只能被插入一个列；RegexHbaseEventSerializer 根据正则表达式将事件的 body 拆分到不同的列中，因此在插入 HBase 的时候，支持用户自定义插入同一个 rowkey 对应的同一个 columnFamily 的多个列。HBaseSink 的优点主要是安全性较高，支持 secure HBase clusters 以及可以向 secure HBase 写入数据（HBase 可以开启 Kerberos 校验）。其缺点主要是性能没有 AsyncHBaseSink 高。因为 HBaseSink 采用了阻塞调用，而 AsyncHBaseSink 采用了非阻塞调用。

HBaseSink 的参数配置如表 3-10 所示。

表 3-10　HBaseSink 的参数配置

参数	默认值	描述
channel	—	为 Agent 的 Channel 命名
type	—	设定类型的名称，如 org.apache.flume.Sink.hbase.HBaseSink
table	—	设置 HBase 的表名
columnFamily	—	设置 HBase 中的 columnFamily
batchSize	100	设定一个事务中处理事件的个数
serializer	org.apache.flume.Sink.hbase.SimpleHbaseEventSerializer; org. apache. flume. Sink. hbase.Regex HbaseEventSerializer	设置 serializer 的处理类
serializer.*	—	要传递给序列化程序的属性（如设置 HBase 的 column）

SimpleHbaseEventSerializer 的配置示例如表 3-11 所示。

表 3-11　SimpleHbaseEventSerializer 的配置示例

配置示例	示例说明
agenttest.channels = c1	为 Agent 的 Channel 命名
agenttest.Sinks = k1	为 Agent 的 Sink 命名
agenttest.Sinks.k1.type = org.apache.flume. Sink.hbase.HBaseSink	设置 type 的值
agenttest.Sinks.k1.table = test_hbase_table	设置 HBase 的 table 的名称

续表

配置示例	示例说明
agenttest.Sinks.k1.columnFamily = f1	设置 HBase 的 columnFamily
agenttest.Sinks.k1.serializer= org.apache.flume.Sink. hbase.SimpleHbaseEventSerializer	设置 serializer 的处理类
agenttest.Sinks.k1.serializer.payloadColumn = columnname	设置 HBase 表的 columnFamily 中的某个列名称
agenttest.Sinks.k1.channels = c1	指定 Sink 使用的 Channel

RegexHbaseEventSerializer 的配置示例如表 3-12 所示。

表 3-12 RegexHbaseEventSerializer 的配置示例

配置示例	示例说明
agenttest.channels = c2	为 Agent 的 Channel 命名
agenttest.Sinks = k2	为 Agent 的 Sink 命名
agenttest.Sinks.k2.type = org.apache.flume.Sink.hbase.HBaseSink	设置 type 的值
agenttest.Sinks.k2.table = test_hbase_table	设置 HBase 的 table 的名称
agenttest.Sinks.k2.columnFamily = f2	设置 HBase 的 columnFamily
agenttest.Sinks.k2.serializer= org.apache.flume.Sink.hbase.RegexHbaseEventSerializer	设置 serializer 的处理类
agenttest.Sinks.k2.serializer.regex = Regular Expression 如：agenttest.Sinks.k2.serializer.regex = \\[(.*?)\\]\\ \\[(.*?)\\]\\ \\[(.*?)\\]\\ \\[(.*?)\\]	设置相应的正则表达式
agent.Sinks.hbaseSink-2.serializer.colNames= column-1,column-2,column-3,column-4	指定前面正则表达式匹配到的数据对应的 HBase 的 familycolumn-2 列族中的 4 个 column 列名
agenttest.Sinks.k2.channels = c2	指定 Sink 使用的 Channel

（2）AsyncHBaseSink

AsyncHBaseSink 目前只提供一种序列化模式：SimpleAsync Hbase EventSerializer。其将整个事件的 body 部分当作完整的一列写入 HBase，因此在插入 HBase 的时候，一个事件的 body 只能被插入一个列。AsyncHBaseSink 的优点主要是其采用非阻塞调用，因此，其性能比 HBaseSink 高；其缺点主要是不支持 secure HBase clusters，且不支持向 secure HBase 写入数据。

AsyncHBaseSink 的参数配置如表 3-13 所示。

表 3-13　AsyncHBaseSink 的参数配置

参数	默认值	描述
channel	—	为 Agent 的 Channel 命名
type	—	设定 type 的名称，此处需要将其设置为 org.apache.flume.Sink.AsyncHBaseSink
table	—	设置 HBase 的表名
columnFamily	—	设置 HBase 中的 columnFamily
batchSize	100	设定一个事务中处理事件的个数
timeout	—	设置 Sink 等待来自 HBase 的 acks 的超时时间（毫秒），如果未设定超时时间，则 Sink 将永远等待
serializer	org.apache.flume.Sink.hbase.SimpleAsyncHbaseEventSerializer	设置 serializer 的处理类
serializer.*	—	要传递给序列化程序的属性（如设置 HBase 的 column）

SimpleAsyncHbaseEventSerializer 的配置示例如表 3-14 所示。

表 3-14　SimpleAsyncHbaseEventSerializer 的配置示例

配置示例	示例说明
agenttest.channels = c3	为 Agent 的 Channel 命名
agenttest.Sinks = k3	为 Agent 的 Sink 命名
agenttest.Sinks.k2.type= org.apache.flume.Sink.hbase.AsyncHBaseSink	设置 type 的值
agenttest.Sinks.k2.table = test_hbase_table	设置 HBase 的 table 的名称
agenttest.Sinks.k2.columnFamily = f3	设置 HBase 的 columnFamily
agenttest.Sinks.k2.serializer = org.apache.flume.Sink.hbase.SimpleAsyncHbaseEventSerializer	设置 serializer 的处理类
agenttest.Sinks.k1.serializer.payloadColumn = columnname	设置 HBase 表的 column Family 中的某个列名称
agenttest.Sinks.k3.channels = c3	指定 Sink 使用的 Channel

3. HDFS 的命令

HDFS 命令的使用方式："hadoop fs -<command> <args>"或"hdfs dfs -<command> <args>"。

HDFS 的常用命令及其含义如下。

（1）help：查看帮助。

（2）ls：查看指定路径的目录结构。

（3）mv：移动或重命名。

（4）rm：删除文件/空白文件夹。

（5）put：上传文件。

（6）mkdir：创建空白文件夹。

任务实施

1. Flume 采集数据上传到 HDFS

进入 Flume 的 conf 目录，创建 hdfs.conf 配置文件，并编辑此配置文件，如图 3-35 所示。

```
# hdfs.conf 配置文件
hdfsAgent.Sources = hdfsSource
hdfsAgent.Sinks = hdfsSinks
hdfsAgent.channels = hdfsChannel
# 配置 Source 信息
hdfsAgent.Sources.hdfsSource.type = spooldir
hdfsAgent.Sources.hdfsSource.channels = hdfsChannel
hdfsAgent.Sources.hdfsSource.spoolDir =
/data/bigdata/apache-flume-1.8.0-bin/20yyxxxxABC/hdfs
hdfsAgent.Sources.hdfsSource.fileHeader = true
# 配置 Sink 信息
hdfsAgent.Sinks.hdfsSinks.type = hdfs
## 文件 hdfs 的存储路径，/%y-%m-%d/%H%M/%S 表示在/年-月-日/时-分/秒目录中'''
hdfsAgent.Sinks.hdfsSinks.hdfs.path=hdfs://master:9000/flume/events/%y-%m-%d/%H%M/%S
# 文件前缀
hdfsAgent.Sinks.hdfsSinks.hdfs.filePrefix=events-
# 文件后缀
hdfsAgent.Sinks.hdfsSinks.hdfs.fileSuffix=log
# 时间取整。注：间隔十分钟以内的会取上一个整时间
hdfsAgent.Sinks.hdfsSinks.hdfs.round=true
hdfsAgent.Sinks.hdfsSinks.hdfs.roundValue=10
hdfsAgent.Sinks.hdfsSinks.hdfs.roundUnit=minute
设置最小备份数为 1，防止因分块造成滚动策略失败而导致生成大量小文件##
hdfsAgent.Sinks.hdfsSinks.hdfs.minBlockReplicas=1
## 设置滚动。注：滚动（roll）指的是，hdfs Sink 将临时文件重命名为最终目标文件，并新打开一个临
时文件来写入数据'''
hdfsAgent.Sinks.hdfsSinks.hdfs.rollInterval=0
hdfsAgent.Sinks.hdfsSinks.hdfs.rollSize=134217728
hdfsAgent.Sinks.hdfsSinks.hdfs.rollCount=0
hdfsAgent.Sinks.hdfsSinks.hdfs.idleTimeout=60
## 设置文件类型
hdfsAgent.Sinks.hdfsSinks.hdfs.fileType=DataStream
## 使用系统内部时间戳
hdfsAgent.Sinks.hdfsSinks.hdfs.useLocalTimeStamp=true
# 配置 Channel 信息
hdfsAgent.channels.hdfsChannel.type = memory
hdfsAgent.channels.hdfsChannel.capacity = 1000
hdfsAgent.channels.hdfsChannel.transactionCapacity = 100
# 将 Source、Sink 绑定到 Channel
hdfsAgent.iSources.hdfsSource.channels = hdfsChannel
hdfsAgent.Sinks.hdfsSinks.channel = hdfsChannel
```

图 3-35 编辑 hdfs.conf 配置文件

使用 HDFS 的命令在 Hadoop 集群根目录中创建/flume/events 文件夹，如图 3-36 所示。

```
hadoop@master:~$ hadoop dfs -mkdir -p /flume/events
DEPRECATED: Use of this script to execute hdfs command is deprecated.
Instead use the hdfs command for it.
```

图 3-36　使用 HDFS 的命令创建文件夹

通过执行命令“flume-ng agent -c. -f /data/bigdata/apache-flume- 1.8.0-bin/conf/hdfs.conf -n hdfsAgent -DFlume.root.logger=INFO,console”启动 flume-ng 监控指定目录，其启动信息如图 3-37 所示。

```
18/08/09 17:38:11 INFO node.Application: Starting new configuration:{ sourceRunners
:{hdfsSource=EventDrivenSourceRunner: { source:Spool Directory source hdfsSource: {
 spoolDir: /data/bigdata/apache-flume-1.8.0-bin/20yyxxxxABC/hdfs } }} sinkRunners:{
hdfsSinks=SinkRunner: { policy:org.apache.flume.sink.DefaultSinkProcessor@35017921
counterGroup:{ name:null counters:{} } }} channels:{hdfsChannel=org.apache.flume.ch
annel.MemoryChannel{name: hdfsChannel}} }
18/08/09 17:38:11 INFO node.Application: Starting Channel hdfsChannel
18/08/09 17:38:11 INFO instrumentation.MonitoredCounterGroup: Monitored counter gro
up for type: CHANNEL, name: hdfsChannel: Successfully registered new MBean.
18/08/09 17:38:11 INFO instrumentation.MonitoredCounterGroup: Component type: CHANN
EL, name: hdfsChannel started
18/08/09 17:38:11 INFO node.Application: Starting Sink hdfsSinks
18/08/09 17:38:11 INFO node.Application: Starting Source hdfsSource
18/08/09 17:38:11 INFO source.SpoolDirectorySource: SpoolDirectorySource source sta
rting with directory: /data/bigdata/apache-flume-1.8.0-bin/20yyxxxxABC/hdfs
18/08/09 17:38:11 INFO instrumentation.MonitoredCounterGroup: Monitored counter gro
up for type: SINK, name: hdfsSinks: Successfully registered new MBean.
18/08/09 17:38:11 INFO instrumentation.MonitoredCounterGroup: Component type: SINK,
 name: hdfsSinks started
18/08/09 17:38:11 INFO instrumentation.MonitoredCounterGroup: Monitored counter gro
up for type: SOURCE, name: hdfsSource: Successfully registered new MBean.
18/08/09 17:38:11 INFO instrumentation.MonitoredCounterGroup: Component type: SOURC
E, name: hdfsSource started
```

图 3-37　flume-ng 的启动信息

打开新的命令行窗口，模拟服务器生成的日志文件,这里模拟生成 4 个日志文件并保存在/data/bigdata/datas/flume-kettle/中，如图 3-38 所示。

```
hadoop@master:/data/bigdata/datas/flume-kettle$ ll
total 24
drwxr-xr-x 2 hadoop root 4096 Aug  9 17:57 ./
drwxr-xr-x 3 hadoop root 4096 Aug  9 17:39 ../
-rw-r--r-- 1 root   root  180 Aug  9 17:56 log.00
-rw-r--r-- 1 root   root  219 Aug  9 17:57 log.01
-rw-r--r-- 1 root   root  212 Aug  9 17:57 log.02
-rw-r--r-- 1 root   root  201 Aug  9 17:57 log.03
```

图 3-38　模拟生成日志文件

执行“cp”命令，将其复制到被监听的文件夹中，模拟服务器生成日志文件，如图 3-39 所示。

```
hadoop@master:/data/bigdata/datas/flume-kettle$ cp ./* /data/bigdata/apache-flum
e-1.8.0-bin/20yyxxxxABC/hdfs/
```

图 3-39　模拟服务器生成日志文件

观察之前运行监控的命令行窗口，发现文件开始复制，文件复制信息如图 3-40 所示。

```
18/08/09 17:58:21 INFO avro.ReliableSpoolingFileEventReader: Preparing to move file
 /data/bigdata/apache-flume-1.8.0-bin/20yyxxxxABC/hdfs/log.00 to /data/bigdata/apac
he-flume-1.8.0-bin/20yyxxxxABC/hdfs/log.00.COMPLETED
18/08/09 17:58:21 INFO hdfs.HDFSDataStream: Serializer = TEXT, UseRawLocalFileSyste
m = false
18/08/09 17:58:21 INFO avro.ReliableSpoolingFileEventReader: Last read took us just
 up to a file boundary. Rolling to the next file, if there is one.
18/08/09 17:58:21 INFO avro.ReliableSpoolingFileEventReader: Preparing to move file
 /data/bigdata/apache-flume-1.8.0-bin/20yyxxxxABC/hdfs/log.01 to /data/bigdata/apac
he-flume-1.8.0-bin/20yyxxxxABC/hdfs/log.01.COMPLETED
18/08/09 17:58:21 INFO avro.ReliableSpoolingFileEventReader: Last read took us just
 up to a file boundary. Rolling to the next file, if there is one.
18/08/09 17:58:21 INFO avro.ReliableSpoolingFileEventReader: Preparing to move file
 /data/bigdata/apache-flume-1.8.0-bin/20yyxxxxABC/hdfs/log.02 to /data/bigdata/apac
he-flume-1.8.0-bin/20yyxxxxABC/hdfs/log.02.COMPLETED
18/08/09 17:58:21 INFO avro.ReliableSpoolingFileEventReader: Last read took us just
 up to a file boundary. Rolling to the next file, if there is one.
18/08/09 17:58:21 INFO avro.ReliableSpoolingFileEventReader: Preparing to move file
 /data/bigdata/apache-flume-1.8.0-bin/20yyxxxxABC/hdfs/log.03 to /data/bigdata/apac
he-flume-1.8.0-bin/20yyxxxxABC/hdfs/log.03.COMPLETED
18/08/09 17:58:21 INFO hdfs.BucketWriter: Creating hdfs://master:9000/flume/events/
18-08-09/1750/00/events-.1533808701382.log.tmp
```

图 3-40 文件复制信息

滚动时间到达 idleTimeout（配置的参数）后才会结束并读出文件，此时，查看文件，如图 3-41 所示。

```
hadoop@master:~$ hadoop -ls -R /flume
Error: No command named `-ls' was found. Perhaps you meant `hadoop ls'
hadoop@master:~$ hadoop dfs  -ls -R /flume
DEPRECATED: Use of this script to execute hdfs command is deprecated.
Instead use the hdfs command for it.

drwxr-xr-x   - hadoop supergroup          0 2018-08-09 17:58 /flume/events
drwxr-xr-x   - hadoop supergroup          0 2018-08-09 17:58 /flume/events/18-08
-09
drwxr-xr-x   - hadoop supergroup          0 2018-08-09 17:58 /flume/events/18-08
-09/1750
drwxr-xr-x   - hadoop supergroup          0 2018-08-09 17:58 /flume/events/18-08
-09/1750/00
-rw-r--r--   3 hadoop supergroup        633 2018-08-09 17:58 /flume/events/18-08
-09/1750/00/events-.1533808701382.log.tmp
hadoop@master:~$ hadoop dfs  -ls -R /flume
DEPRECATED: Use of this script to execute hdfs command is deprecated.
Instead use the hdfs command for it.

drwxr-xr-x   - hadoop supergroup          0 2018-08-09 17:58 /flume/events
drwxr-xr-x   - hadoop supergroup          0 2018-08-09 17:58 /flume/events/18-08
-09
drwxr-xr-x   - hadoop supergroup          0 2018-08-09 17:58 /flume/events/18-08
-09/1750
drwxr-xr-x   - hadoop supergroup          0 2018-08-09 17:59 /flume/events/18-08
-09/1750/00
-rw-r--r--   3 hadoop supergroup        633 2018-08-09 17:59 /flume/events/18-08
-09/1750/00/events-.1533808701382.log
```

图 3-41 查看文件

企业实际生成环境一般会根据自身日志量和日志特点指定文件滚动规则。这里由于实例数据量较少，所以采用大小 128MB、时间 60s 作为文件滚动规则。

2. Flume 采集数据上传到 HBase

（1）利用 SimpleHbaseEventSerializer 序列化模式上传数据。

首先，在 HBase 中创建一个表 mikeal-hbase-table，表中拥有 familyclom1 和 familyclom2 两个列族，如图 3-42 所示。

```
hbase(main):004:0> create 'mikeal-hbase-table','familyclom1','familyclom2'
0 row(s) in 2.4430 seconds

=> Hbase::Table - mikeal-hbase-table
```

图 3-42 在 HBase 中创建表

在 Flume 安装目录的 conf 目录中创建并配置 test-flume-into-hbase.conf 文件，如图 3-43 所示。

```
#从文件中读取实时信息，不做处理直接存储到HBase中
agent.sources  = logfile-source
agent.channels  = file-channel
agent.sinks = hbase-sink

#logfile-source配置
agent.sources.logfile-source.type = exec
agent.sources.logfile-source.command= tail -f /data/flume-hbase-test/mkhbasetable/data/test.log
agent.sources.logfile-source.checkerpiodic = 50

#组合Source和Channel
agent.sources.logfile-source.channels = file-channel

#配置Channel,使用本地file
agent.channels.file-channel.type = file
agent.channels.file-channel.checkpointDir = /data/flume-hbase-test/checkpoint
agent.channels.file-channel.dataDirs = /data/flume-hbase-test/data

#将Sink配置为HBaseSink和simpleHbaseEventSerializer
agent.sinks.hbase-sink.type = org.apache.flume.sink.hbase.HBaseSink
#HBase表名
agent.sinks.hbase-sink.table = mikeal-hbase-table

#HBase表的列族名称
agent.sinks.hbase-sink.columnFamily = familyclom1
agent.sinks.hbase-sink.serializer = org.apache.flume.sink.hbase.simpleHbaseEventSerializer

#HBase表的列族中的某个列名称
agent.sinks.hbase-sink.serializer.payloadcolumn = column-1

#组合Sink和Channel
agent.sinks.hbase-sink.channel = file-channel
```

图 3-43 配置 test-flume-into-hbase.conf 文件

在配置文件中，选择本地的/data/flume-hbase-test/mkhbasetable/data/test.log 文件作为实时数据采集源日志文件，选择本地文件目录/data/flume-hbase-test/data 作为 Channel，选择 HBase 为 Sink（即数据流写入 HBase）。

在 Flume 安装目录的 bin 目录中启动 Flume，如图 3-44 所示。

```
hadoop@ubuntu:/usr/local/flume/apache-flume-1.8.0-bin/bin$ ./flume-ng agent
--name agent -f /usr/local/flume/apache-flume-1.8.0-bin/conf/test-flume-into
-hbase.conf -c ../conf/ -Dflume.root.logger=INFO,console
```

图 3-44　启动 Flume（1）

重新打开一个命令行窗口，向 test.log 文件中输入测试数据，如图 3-45 所示。需要提前准备好测试用日志文件“数据.txt”。

```
hadoop@ubuntu:/data/flume-hbase-test/mkhbasetable/data$ cat 数据.txt >> test.log
```

图 3-45　向 test.log 文件中输入测试数据（1）

查看 mikeal-hbase-table，其显示的内容正是向 test.log 文件中传入的数据，如图 3-46 所示。

```
hbase(main):021:0> scan 'mikeal-hbase-table'
ROW                         COLUMN+CELL
 default0c75ef3c-0886-4     column=familyclom1:cloumn-1, timestamp=1532938323385, value=124.2
 61c-926b-f514f241e37b      39.208.72\x0920130530184853\x09favicon.ico
 default11e35a36-6181-4     column=familyclom1:cloumn-1, timestamp=1532938323385, value=110.9
 617-bf38-cf5cd8476e94      0.190.163\x0920130531020232\x09forum.php
 default12380865-205e-4     column=familyclom1:cloumn-1, timestamp=1532938323385, value=124.2
 8c1-83bb-b83d0c215732      28.0.205\x0920130531020226\x09home.php?mod=spacecp&ac=pm&op=check
                            newpm&rand=1369936943
 default134f18f4-0d10-4     column=familyclom1:cloumn-1, timestamp=1532938326284, value=61.13
 7b3-a56b-95d6402bb869      5.249.219\x0920130531020250\x09forum.php?action=postreview&do=sup
                            port&hash=21eda218&mod=misc&pid=46916&tid=10682
 default13a08436-6884-4     column=familyclom1:cloumn-1, timestamp=1532938326284, value=168.6
 688-8b1a-48ca96d410d4      3.121.23\x0920130531020250\x09api.php?mod=js&bid=65
 default1495fbaf-56d0-4     column=familyclom1:cloumn-1, timestamp=1532938323385, value=110.1
 d50-beda-71855e7c6b45      78.201.232\x0920130531020226\x09data/attachment/common/c2/common_
                            12_usergroup_icon.jpg
 default150cbb0b-7d6a-4     column=familyclom1:cloumn-1, timestamp=1532938323385, value=110.1
 70c-9d44-0c2bb86be68d      79.82.61\x0920130530184856\x09api.php?mod=js&bid=67
 default15656703-afe3-4     column=familyclom1:cloumn-1, timestamp=1532938323385, value=203.2
 4a4-897f-6cdd40eaaa43      08.60.92\x0920130530184856\x09home.php?mod=space&uid=55635&do=blo
                            g&view=me
 default18712078-d16a-4     column=familyclom1:cloumn-1, timestamp=1532938326284, value=110.9
 3fc-a55d-e71d82688cf6      0.190.163\x0920130531020247\x09home.php?mod=spacecp&ac=profile&op
```

图 3-46　查看 mikeal-hbase-table（1）

（2）利用 SimpleAsyncHbaseEventSerializer 序列化模式上传数据。

为了使示例更加清晰，先把 mikeal-hbase-table 数据清空，执行如下命令。

```
truncate 'mikeal-hbase-table'
```

在 Flume 安装目录的 conf 目录中创建并配置 test-flume-into-hbase-2.conf 文件，如图 3-47 所示。

在 Flume 安装目录的 bin 目录中启动 Flume，如图 3-48 所示。

重新打开一个命令行窗口，向 test.log 文件中输入测试数据，如图 3-49 所示。需要提前准备好测试用日志文件“数据.txt”。

再次查看 mikeal-hbase-table，其显示的内容正是向 test.log 文件中传入的数据，如图 3-50 所示。

```
# 从文件中读取实时信息，不作处理直接存储到HBase中
agent.sources = logfile-source
agent.channels = file-channel
agent.sinks = hbase-sink

# logfile-source配置
agent.sources.logfile-source.type = exec
agent.sources.logfile-source.command = tail -f /data/flume-hbase-test/mkhbas
etable/data/test.log
agent.sources.logfile-source.checkperiodic = 50

# 配置Channel，使用本地file
agent.channels.file-channel.type = file
agent.channels.file-channel.checkpointDir = /data/flume-hbase-test/checkpoin
t
agent.channels.file-channel.dataDirs = /data/flume-hbase-test/data

#将Sink配置为HBase
agent.sinks.hbase-sink.type = org.apache.flume.sink.hbase.AsyncHBaseSink
agent.sinks.hbase-sink.table = mikeal-hbase-table
agent.sinks.hbase-sink.columnFamily  = familyclom1
agent.sinks.hbase-sink.serializer = org.apache.flume.sink.hbase.SimpleAsyncH
baseEventSerializer
agent.sinks.hbase-sink.serializer.payloadColumn = cloumn-1

# 组合Source、Sink和Channel
agent.sources.logfile-source.channels = file-channel
agent.sinks.hbase-sink.channel = file-channel
```

图 3-47　配置 test-flume-into-hbase-2.conf 文件

```
hadoop@ubuntu:/usr/local/flume/apache-flume-1.8.0-bin/bin$ ./flume-ng agent -n
agent -f /usr/local/flume/apache-flume-1.8.0-bin/conf/test-flume-into-hbase-2.conf
-c ../conf/ -Dflume.root.logger=INFO,console
```

图 3-48　启动 Flume（2）

```
hadoop@ubuntu:/data/flume-hbase-test/mkhbasetable/data$ cat 数据.txt >> test.log
```

图 3-49　向 test.log 文件中输入测试数据（2）

```
hbase(main):028:0> scan 'mikeal-hbase-table'
ROW                   COLUMN+CELL
 default0025526b-79a0-4 column=familyclom1:cloumn-1, timestamp=1532945521001, value=222.
 082-8c68-39be729e3ac0  82.221.214\x0920130530184854\x09api.php?mod=js&bid=66
 default0682f205-d62b-4 column=familyclom1:cloumn-1, timestamp=1532945520501, value=222.
 0a9-a18d-b5653834a47e  133.189.179\x0920130531020232\x09data/cache/style_1_common.css?y
                        7a
 default07338a83-3d04-4 column=familyclom1:cloumn-1, timestamp=1532945520068, value=1.20
 3ae-997a-d9f2c9b30369  2.219.124\x0920130531020223\x09home.php?mod=space&uid=12026&do=s
                        hare&view=me&from=space&type=thread
 default0c7d93a1-594c-4 column=familyclom1:cloumn-1, timestamp=1532945520588, value=222.
 ad8-956f-c8a230b79072  133.189.179\x0920130531020232\x09source/plugin/dsu_kkvip/images/
                        vip.png
 default0fe3cc51-65de-4 column=familyclom1:cloumn-1, timestamp=1532945520438, value=110.
 f96-a78c-a092d67bd1d3  90.190.163\x0920130531020230\x09home.php?mod=spacecp&ac=profile&
                        op=info
 default15dbceb0-98c1-4 column=familyclom1:cloumn-1, timestamp=1532945520897, value=14.1
 67f-b5d2-b015a849184c  54.192.149\x0920130531020240\x09forum.php?mod=post&action=newthr
                        ead&fid=114&extra=&topicsubmit=yes&ajaxmenu=1&inajax=1
 default173690bd-e9e9-4 column=familyclom1:cloumn-1, timestamp=1532945519806, value=124.
 60b-a64d-5a4e13510c8c  239.208.72\x0920130530184851\x09api/connect/like.php
 default19d3f0be-501e-4 column=familyclom1:cloumn-1, timestamp=1532945520398, value=23.2
 2b6-be42-c18e0757d53d  2.64.87\x0920130531020231\x09robots.txt
 default1bd40544-1f3b-4 column=familyclom1:cloumn-1, timestamp=1532945520398, value=173.
 10d-9eb5-464e5deeca57  236.32.218\x0920130531020230\x09 HTTP/1.0
 default21c2bdba-0b20-4 column=familyclom1:cloumn-1, timestamp=1532945520949, value=123.
 fac-b655-dad1739b709c  125.71.114\x0920130531020242\x09?29690
 default240187ce-a79c-4 column=familyclom1:cloumn-1, timestamp=1532945520655, value=110.
 2a4-b7e7-77c14094d3cf  90.190.163\x0920130531020236\x09member.php?mod=logging&action=lo
```

图 3-50　查看 mikeal-hbase-table（2）

（3）利用 RegexHbaseEventSerializer 序列化模式上传数据。

RegexHbaseEventSerializer 可以使用正则表达式匹配切割事件，并存入 HBase 表的多个列。

为了使示例更加清晰，先将 mikeal-hbase-table 数据清空，执行如下命令。

```
truncate 'mikeal-hbase-table'
```

在 Flume 安装目录的 conf 目录中创建并配置 test-flume-into-hbase-3.conf 文件，如图 3-51 所示。

```
# 从文件中读取实时信息，不做处理直接存储到HBase中
agent.sources = logfile-source
agent.channels = file-channel
agent.sinks = hbase-sink

# logfile-source配置
agent.sources.logfile-source.type = exec
agent.sources.logfile-source.command = tail -f /data/flume-hbase-test/mkhbas
etable/data/nginx.log
agent.sources.logfile-source.checkperiodic = 50

# 配置Channel，使用本地file
agent.channels.file-channel.type = file
agent.channels.file-channel.checkpointDir = /data/flume-hbase-test/checkpoin
t
agent.channels.file-channel.dataDirs = /data/flume-hbase-test/data

# 将Sink配置为HBase
agent.sinks.hbase-sink.type = org.apache.flume.sink.hbase.HBaseSink
agent.sinks.hbase-sink.table = mikeal-hbase-table
agent.sinks.hbase-sink.columnFamily  = familyclom1
agent.sinks.hbase-sink.serializer = org.apache.flume.sink.hbase.RegexHbaseEv
entSerializer
# 对test日志做分割，按列存储到HBase中，正则匹配分成的列为 (.*?)  (.*?)
(.*?)
agent.sinks.hbase-sink.serializer.regex =(.*?)  (.*?)   (.*?)
agent.sinks.hbase-sink.serializer.colNames = ip, time, url

# 组合Source、Sink和Channel
agent.sources.logfile-source.channels = file-channel
agent.sinks.hbase-sink.channel = file-channel
```

图 3-51　配置 test-flume-into-hbase-3.conf 文件

在 Flume 安装目录的 bin 目录中启动 Flume，如图 3-52 所示。

```
hadoop@ubuntu:/usr/local/flume/apache-flume-1.8.0-bin/bin$ ./flume-ng agent -n a
gent -f /usr/local/flume/apache-flume-1.8.0-bin/conf/test-flume-into-hbase-3.con
f -c ../conf/ -Dflume.root.logger=INFO,console
```

图 3-52　启动 Flume（3）

重新打开一个命令行窗口，向 test.log 文件中输入测试数据，如图 3-53 所示。需要提前准备好测试用日志文件“数据.txt”。

```
hadoop@ubuntu:/data/flume-hbase-test/mkhbasetable/data$ cat 数据.txt >> test.log
```

图 3-53　向 test.log 文件中输入测试数据（3）

查看 mikeal-hbase-table，其显示的内容正是向 test.log 文件中传入的数据，如图 3-54 所示。

（4）多 Source、多 Channel 和多 Sink 的案例实现。

为了使示例更加清晰，执行“truncate 'mikeal-hbase-table'”将 mikeal-hbase-table 数据清空。

在 Flume 安装目录的 conf 目录中创建并配置 test-flume-into-hbase-multi-position.conf 文件，如图 3-55 所示。

```
hbase(main):026:0> scan 'mikeal-hbase-table'
ROW                          COLUMN+CELL
 1532943839244-Izwnkcgn      column=familyclom1: time, timestamp=1532943839784, value=20130531
 ge-0                        020251
 1532943839244-Izwnkcgn      column=familyclom1: url, timestamp=1532943839784, value=forum-66-
 ge-0                        1.html
 1532943839244-Izwnkcgn      column=familyclom1:ip, timestamp=1532943839784, value=110.90.190.
 ge-0                        163
 1532943839258-Izwnkcgn      column=familyclom1: time, timestamp=1532943839784, value=20130530
 ge-1                        184851
 1532943839258-Izwnkcgn      column=familyclom1: url, timestamp=1532943839784, value=forum.php
 ge-1                        ?mod=image&aid=18322&size=300x300&key=6c372c86470a6cea&nocache=ye
                             s&type=fixnone
 1532943839258-Izwnkcgn      column=familyclom1:ip, timestamp=1532943839784, value=\xEF\xBB\xB
 ge-1                        F\xEF\xBB\xBF111.145.51.149
 1532943839259-Izwnkcgn      column=familyclom1: time, timestamp=1532943839784, value=20130531
 ge-2                        020223
 1532943839259-Izwnkcgn      column=familyclom1: url, timestamp=1532943839784, value=forum.php
 ge-2
 1532943839259-Izwnkcgn      column=familyclom1:ip, timestamp=1532943839784, value=220.181.108
 ge-2                        .89
 1532943839260-Izwnkcgn      column=familyclom1: time, timestamp=1532943839784, value=20130531
 ge-3                        020223
```

图 3-54　查看 mikeal-hbase-table（3）

```
# 从文件中读取实时信息，不做处理直接存储到HBase中
agent.sources = logfile-source-1 logfile-source-2
agent.channels = file-channel-1 file-channel-2
agent.sinks = hbase-sink-1 hbase-sink-2

# logfile-source配置
agent.sources.logfile-source-1.type = exec
agent.sources.logfile-source-1.command = tail -f /data/flume-hbase-test/mkhb
asetable/data/nginx.log
agent.sources.logfile-source-1.checkperiodic = 50

agent.sources.logfile-source-2.type = exec
agent.sources.logfile-source-2.command = tail -f /data/flume-hbase-test/mkhb
asetable/data/tomcat.log
agent.sources.logfile-source-2.checkperiodic = 50

# 配置Channel，使用本地file
agent.channels.file-channel-1.type = file
agent.channels.file-channel-1.checkpointDir = /data/flume-hbase-test/checkpo
int
agent.channels.file-channel-1.dataDirs = /data/flume-hbase-test/data

agent.channels.file-channel-2.type = file
agent.channels.file-channel-2.checkpointDir = /data/flume-hbase-test/checkpo
int2
agent.channels.file-channel-2.dataDirs = /data/flume-hbase-test/data2

# 将Sink配置为HBase
agent.sinks.hbase-sink-1.type = org.apache.flume.sink.hbase.HBaseSink
agent.sinks.hbase-sink-1.table = mikeal-hbase-table
agent.sinks.hbase-sink-1.columnFamily  = familyclom1
agent.sinks.hbase-sink-1.serializer = org.apache.flume.sink.hbase.RegexHbase
EventSerializer
# 对test日志做切割，并按列存储HBase
agent.sinks.hbase-sink-1.serializer.regex =(.*?)    (.*?)   (.*?)
agent.sinks.hbase-sink-1.serializer.colNames = ip,time,url
agent.sinks.hbase-sink-2.type = org.apache.flume.sink.hbase.HBaseSink
agent.sinks.hbase-sink-2.table = mikeal-hbase-table
agent.sinks.hbase-sink-2.columnFamily  = familyclom2
agent.sinks.hbase-sink-2.serializer = org.apache.flume.sink.hbase.RegexHbase
EventSerializer
agent.sinks.hbase-sink-2.serializer.regex =(.*?)    (.*?)   (.*?)
agent.sinks.hbase-sink-2.serializer.colNames = ip,time,url

# 组合Source、Sink和Channel
agent.sources.logfile-source-1.channels = file-channel-1
agent.sinks.hbase-sink-1.channel = file-channel-1

agent.sources.logfile-source-2.channels = file-channel-2
agent.sinks.hbase-sink-2.channel = file-channel-2
```

图 3-55　配置 test-flume-into-hbase-multi-position.conf 文件

在 Flume 安装目录的 bin 目录中启动 Flume，如图 3-56 所示。

```
hadoop@ubuntu:/usr/local/flume/apache-flume-1.8.0-bin/bin$ ./flume-ng agent -n a
gent -f /usr/local/flume/apache-flume-1.8.0-bin/conf/test-flume-into-hbase-multi
-position.conf -c ../conf/ -Dflume.root.logger=INFO,console
```

图 3-56　启动 Flume（4）

重新打开一个命令行窗口，分别向 test.log 和 test2.log 文件中输入测试数据，如图 3-57 所示。需要提前准备好测试用日志文件“数据.txt”。

```
hadoop@ubuntu:/data/flume-hbase-test/mkhbasetable/data$ cat 数据.txt >> test.log
hadoop@ubuntu:/data/flume-hbase-test/mkhbasetable/data$ cat 数据.txt >> test2.log
```

图 3-57　输入测试数据

查看 mikeal-hbase-table，其显示的内容正是向 test.log 和 test2.log 文件中传入的数据，如图 3-58 所示。

```
hbase(main):014:0> scan 'mikeal-hbase-table'
ROW                    COLUMN+CELL
 1533032155256-aZJLEz column=familyclom1:ip, timestamp=1533032160183, value=110.
 gzVv-0               90.190.163
 1533032155256-aZJLEz column=familyclom1:time, timestamp=1533032160183, value=20
 gzVv-0               130531020251
 1533032155256-aZJLEz column=familyclom1:url, timestamp=1533032160183, value=for
 gzVv-0               um-66-1.html\xEF\xBB\xBF
 1533032155262-aZJLEz column=familyclom2:ip, timestamp=1533032160702, value=110.
 gzVv-1               90.190.163
 1533032155262-aZJLEz column=familyclom2:time, timestamp=1533032160702, value=20
 gzVv-1               130531020251
 1533032155262-aZJLEz column=familyclom2:url, timestamp=1533032160702, value=for
 gzVv-1               um-66-1.html
 1533032155339-aZJLEz column=familyclom1:ip, timestamp=1533032160183, value=59.1
 gzVv-2               26.195.102
 1533032155339-aZJLEz column=familyclom1:time, timestamp=1533032160183, value=20
 gzVv-2               130531020247
 1533032155339-aZJLEz column=familyclom1:url, timestamp=1533032160183, value=api
 gzVv-2               .php?mod=js&bid=94
 1533032155340-aZJLEz column=familyclom1:ip, timestamp=1533032160183, value=110.
 gzVv-3               90.190.163
```

图 3-58　查看 mikeal-hbase-table（4）

任务 3　创新与拓展

任务描述

（1）使用 Flume 采集 Linux 系统日志。

（2）使用 Flume 采集 Nginx 日志。

（3）使用 Kettle 采集 Excel 表格中的数据并导入 MySQL 数据库。

（4）使用 Kettle 采集 MySQL 数据库和 Excel 表格中的数据，将其导入 MongoDB 数据库并进行关联。

任务目标

（1）了解 Flume 日志信息的采集过程。

（2）熟悉 Nginx 的安装及其基础语法。

（3）学会使用 Ubuntu 和 Flume 搭建日志采集系统，并完成日志采集。

（4）学会使用 Nginx 和 Flume 搭建日志采集系统，并完成日志采集。

（5）学会使用 Kettle 采集 Excel 表格中的数据。

（6）学会使用 Kettle 进行多复合来源数据的采集与关联。

项目 4 数据预处理实践

04

学习目标

【知识目标】

了解大数据预处理工具的使用方法。

【技能目标】

① 学会使用 Pig 进行数据预处理。
② 学会使用 Kettle 进行数据预处理。
③ 学会使用 Pandas 进行数据预处理。
④ 学会使用 OpenRefine 进行数据预处理。
⑤ 学会使用 Flume Interceptor 进行日志数据预处理。

项目描述

数据预处理是大数据领域不可缺少的环节，用来发现并纠正数据中可能存在的错误，针对数据审查过程中发现的错误值、缺失值、异常值、可疑数据，选用适当方法进行“清理”，使“脏”数据变为“干净”数据。在大数据处理过程中，数据预处理将占用 60%～80%的时间。

本项目主要使用常见的工具（Pig、Kettle、Pandas、OpenRefine、Flume）完成数据预处理。

任务 1 用 Pig 进行数据预处理

任务描述

（1）学习 Pig 的相关基础知识。

（2）使用 Pig 实现“北京公交线路信息”数据的预处理。

任务目标

（1）熟悉 Pig 的相关基础知识。

（2）学会使用 Pig 完成“北京公交线路信息”数据的预处理。

知识准备

下面来了解 Pig 中主要的操作。

在本地文件系统中，创建一个包含数据的输入文件 student_data.txt，并执行“put”命令，将文件从本地文件系统上传到 HDFS 中。student_data.txt 文件内容如下所示。

```
001,Rajiv,Reddy,9848022337,Hyderabad
002,Siddarth,Battacharya,9848022338,Kolkata
003,Rajesh,Khanna,9848022339,Delhi
004,Preethi,Agarwal,9848022330,Pune
005,Trupthi,Mohanthy,9848022336,Bhuwaneshwar
006,Archana,Mishra,9848022335,Chennai
```

（1）载入和存储

① 载入：从文件系统或其他存储介质中载入数据到一个 relation 中。

载入语句由两部分组成，由“=”运算符分隔，其左侧为存储数据关系的名称，右侧定义如何存储数据。下面给出了 LOAD 运算符的语法。

```
Relation_name = LOAD 'Input file path' USING function as schema;
```

其中，各参数的说明如下。

a. Relation_name：该参数必须明确给出，推荐命名明确体现存储数据的关系。

b. Input file path：该参数必须明确给出，其用于指定存储文件的 HDFS 目录（在 MapReduce 模式下）。

c. function：该参数必须明确给出，其需要从 Apache Pig 提供的一组加载函数中选择一个函数（BinStorage、JsonLoader、PigStorage、TextLoader）。

d. schema：该参数必须明确给出，其用于定义数据的模式，如 column1 : data type, column2 : data type, column3 : data type。

现在，通过在 Grunt Shell 中执行以下 Pig Latin 语句，将 student_data.txt 文件中的数据加载到 Pig 中。

```
grunt> student = LOAD 'hdfs://localhost:9000/pig_data/student_data.txt' USING PigStorage(',') as ( id:int, firstname:chararray, lastname:chararray, phone:chararray, city:chararray );
```

其参数说明如表 4-1 所示。

表 4-1　参数说明

参数	说明
Relation name	已将数据存储在学生（student）模式中
Input file path	从 HDFS 的/pig_data/目录的 student_data.txt 文件中读取数据
Storage function	使用了 PigStorage()函数，将数据加载并存储为结构化文本文件。函数采用了分隔符','，默认情况下，函数以“\t”作为参数
schema	例如，使用以下模式存储数据： column: id, firstname, lastname, phone, city datatype: int, chararray, chararray, chararray, chararray

column	id	firstname	lastname	phone	city
datatype	int	chararray	chararray	chararray	chararray

注：Load 语句会简单地将数据加载到 Pig 指定的关系中。

② 存储：保存 relation 到文件系统或其他存储介质中。

下面给出了 STORE 运算符的语法。

```
STORE Relation_name INTO ' required_directory_path ' [USING function];
```

使用 LOAD 运算符将其读入关系 student，代码如下。

```
grunt> student = LOAD 'hdfs://localhost:9000/pig_data/student_data.txt' USING
PigStorage(',') as ( id:int, firstname:chararray, lastname:chararray, phone:chararray,
city:chararray );
```

将关系存储在 HDFS 目录“/pig_Output/”中，代码如下。

```
grunt> STORE student INTO ' hdfs://localhost:9000/pig_Output/ ' USING PigStorage (',');
```

执行 STORE 语句后，将使用指定的名称创建目录，并将数据存储在其中。

③ 备份：输出一个 relation 到控制台，并在控制台上输出关系的内容。

（2）过滤

① FILTER 运算符。

FILTER 运算符用于根据条件从关系中选择所需的元组。下面给出了 FILTER 运算符的语法。

```
grunt> Relation2_name = FILTER Relation1_name BY (condition);
```

在 HDFS 目录/pig_data/中有一个名称为 student_details.txt 的文件，文件内容如下所示。

```
001,Rajiv,Reddy,21,9848022337,Hyderabad
002,Siddarth,Battacharya,22,9848022338,Kolkata
003,Rajesh,Khanna,22,9848022339,Delhi
004,Preethi,Agarwal,21,9848022330,Pune
005,Trupthi,Mohanthy,23,9848022336,Bhuwaneshwar
006,Archana,Mishra,23,9848022335,Chennai
007,Komal,Nayak,24,9848022334,Trivendram
008,Bharathi,Nambiayar,24,9848022333,Chennai
```

将此文件通过关系 student_details 加载到 Pig 中，代码如下。

```
grunt> student_details = LOAD 'hdfs://localhost:9000/pig_data/student_details.txt'
USING PigStorage(',') as (id:int, firstname:chararray, lastname:chararray, age:int,
phone:chararray, city:chararray);
```

使用 FILTER 运算符获取属于 Chennai 城市的学生的详细信息，代码如下。

```
filter_data = FILTER student_details BY city == 'Chennai';
```

使用“DUMP filter_data;”输出如下对应关系的信息。

```
(6,Archana,Mishra,23,9848022335,Chennai)
(8,Bharathi,Nambiayar,24,9848022333,Chennai)
```

② DISTINCT 运算符。

DISTINCT 运算符用于从关系中删除冗余（重复）元组。下面给出了 DISTINCT 运算符的语法。

```
grunt> Relation_name2 = DISTINCT Relation_name1;
```

在 HDFS 目录/pig_data/中有一个名称为 student_details.txt 的文件，文件内容如下所示。

```
001,Rajiv,Reddy,9848022337,Hyderabad
002,Siddarth,Battacharya,9848022338,Kolkata
002,Siddarth,Battacharya,9848022338,Kolkata
003,Rajesh,Khanna,9848022339,Delhi
003,Rajesh,Khanna,9848022339,Delhi
004,Preethi,Agarwal,9848022330,Pune
```

```
005,Trupthi,Mohanthy,9848022336,Bhuwaneshwar
006,Archana,Mishra,9848022335,Chennai
006,Archana,Mishra,9848022335,Chennai
```

通过关系 student_details 将此文件加载到 Pig 中，代码如下。

```
grunt> student_details = LOAD 'hdfs://localhost:9000/pig_data/student_details.txt' USING PigStorage(',') as (id:int, firstname:chararray, lastname:chararray, phone:chararray, city:chararray);
```

使用 DISTINCT 运算符从 student_details 关系中删除冗余（重复）元组，并将其另存在 distinct_data 的关系中，代码如下。

```
grunt> distinct_data = DISTINCT student_details;
```

执行“DUMP distinct_data;”命令，输出如下对应关系的信息，在输出的信息中可见重复的元组已被删除。

```
(1,Rajiv,Reddy,9848022337,Hyderabad)
(2,Siddarth,Battacharya,9848022338,Kolkata)
(3,Rajesh,Khanna,9848022339,Delhi)
(4,Preethi,Agarwal,9848022330,Pune)
(5,Trupthi,Mohanthy,9848022336,Bhuwaneshwar)
(6,Archana,Mishra,9848022335,Chennai)
```

③ FOREACH 运算符。

FOREACH 运算符用于基于列数据生成指定的数据转换。下面给出了 FOREACH 运算符的语法。

```
grunt> Relation_name2 = FOREACH Relation_name1 GENERATE (required data);
```

在 HDFS 目录/pig_data/中有一个名称为 student_details.txt 的文件，文件内容如下所示。

```
001,Rajiv,Reddy,21,9848022337,Hyderabad
002,Siddarth,Battacharya,22,9848022338,Kolkata
003,Rajesh,Khanna,22,9848022339,Delhi
004,Preethi,Agarwal,21,9848022330,Pune
005,Trupthi,Mohanthy,23,9848022336,Bhuwaneshwar
006,Archana,Mishra,23,9848022335,Chennai
007,Komal,Nayak,24,9848022334,Trivendram
008,Bharathi,Nambiayar,24,9848022333,Chennai
```

通过关系 student_details 将此文件加载到 Pig 中，代码如下。

```
grunt> student_details = LOAD 'hdfs://localhost:9000/pig_data/student_details.txt' USING PigStorage(',') as (id:int, firstname:chararray, lastname:chararray,age:int, phone:chararray, city:chararray);
```

从关系 student_details 中获取每个学生的 id、age 和 city，并使用 FOREACH 运算符将其存储到 foreach_data 关系中，代码如下。

```
grunt> foreach_data = FOREACH student_details GENERATE id,age,city;
```

执行“DUMP foreach_data;”命令，输出如下对应关系的信息。

```
(1,21,Hyderabad)
(2,22,Kolkata)
(3,22,Delhi)
(4,21,Pune)
(5,23,Bhuwaneshwar)
(6,23,Chennai)
(7,24,Trivendram)
(8,24,Chennai)
```

（3）分组与关联

① JOIN 运算符。

JOIN 运算符用于连接来自两个或多个关系的记录。在执行连接操作时，从每个关系中声明一个（或一组）元组作为 key。当这些 key 匹配时，两个特定的元组匹配，否则记录将被丢弃。连接可以为以下类型：Self-join、Inner-join、Outer-join（包括左外连接、右外连接及全外连接）。

在 HDFS 的/pig_data/目录中有两个文件，即 customers.txt 和 orders.txt，其中 customers.txt 内容如下所示。

```
1,Ramesh,32,Ahmedabad,2000.00
2,Khilan,25,Delhi,1500.00
3,Kaushik,23,Kota,2000.00
4,Chaitali,25,Mumbai,6500.00
5,Hardik,27,Bhopal,8500.00
6,Komal,22,MP,4500.00
7,Muffy,24,Indore,10000.00
```

orders.txt 内容如下所示。

```
102,2009-10-08 00:00:00,3,3000
100,2009-10-08 00:00:00,3,1500
101,2009-11-20 00:00:00,2,1560
103,2008-05-20 00:00:00,4,2060
```

将这两个文件加载到 Pig 中，代码如下。

```
grunt> customers = LOAD 'hdfs://localhost:9000/pig_data/customers.txt' USING PigStorage(',') as (id:int, name:chararray, age:int, address:chararray, salary:int);

grunt> orders = LOAD 'hdfs://localhost:9000/pig_data/orders.txt' USING PigStorage(',') as (oid:int, date:chararray, customer_id:int, amount:int);
```

对这两个关系执行各种连接操作，各操作如下所示。

a. Self-join（自连接）。

其语法格式如下。

```
grunt> Relation3_name = JOIN Relation1_name BY key, Relation2_name BY key ;
```

Self-join 用于将表与其自身连接，就像两个关系进行连接一样，需要临时重命名至少一个关系。通常，在 Apache Pig 中，为了执行自连接操作，将在不同的别名（名称）下多次加载相同的数据。这里要将文件 customers.txt 的内容加载到两个不同命名的关系中，代码如下。

```
grunt> customers1 = LOAD 'hdfs://localhost:9000/pig_data/customers.txt' USING PigStorage(',') as (id:int, name:chararray, age:int, address:chararray, salary:int);

grunt> customers2 = LOAD 'hdfs://localhost:9000/pig_data/customers.txt' USING PigStorage(',') as (id:int, name:chararray, age:int, address:chararray, salary:int);
```

通过两个关系 customers1 和 customers2，对关系 customers 执行自连接操作，代码如下。

```
grunt> customers3 = JOIN customers1 BY id, customers2 BY id;
```

执行“DUMP customers3;”命令，输出对应关系的信息，如下所示。

```
(1,Ramesh,32,Ahmedabad,2000,1,Ramesh,32,Ahmedabad,2000)
(2,Khilan,25,Delhi,1500,2,Khilan,25,Delhi,1500)
(3,kaushik,23,Kota,2000,3,Kaushik,23,Kota,2000)
(4,Chaitali,25,Mumbai,6500,4,Chaitali,25,Mumbai,6500)
(5,Hardik,27,Bhopal,8500,5,Hardik,27,Bhopal,8500)
```

```
(6,Komal,22,MP,4500,6,Komal,22,MP,4500)
(7,Muffy,24,Indore,10000,7,Muffy,24,Indore,10000)
```

b. Inner-join（内部连接）。

Inner-join 使用较为频繁，也被称为等值连接。当两个表中都存在匹配时，内部连接将返回行。基于连接谓词，通过组合两个关系（如 A 和 B）的列值来创建新关系。对 A 的每一行与 B 的每一行进行比较，以查找满足连接谓词的所有行对。当连接谓词被满足时，A 和 B 的每个匹配的行对的列值被组合成结果行。其语法格式如下。

```
grunt> result = JOIN relation1 BY columnname, relation2 BY columnname;
```

对 customers 和 orders 执行内部连接操作，代码如下。

```
grunt> customer_orders = JOIN customers BY id, orders BY customer_id;
```

执行“DUMP customer_orders;”命令，输出如下对应关系的信息。

```
(2,Khilan,25,Delhi,1500,101,2009-11-20 00:00:00,2,1560)
(3,Kaushik,23,Kota,2000,100,2009-10-08 00:00:00,3,1500)
(3,Kaushik,23,Kota,2000,102,2009-10-08 00:00:00,3,3000)
(4,Chaitali,25,Mumbai,6500,103,2008-05-20 00:00:00,4,2060)
```

c. Outer-join（外部连接）。

外部连接与内部连接不同，外部连接返回至少一个关系中的所有行。外部连接操作以 3 种方式执行：左外连接、右外连接和全外连接。

ⓐ 左外连接：左外连接返回左表中的所有行，即使右边的关系中没有匹配项。其语法格式如下。

```
grunt> Relation3_name = JOIN Relation1_name BY id LEFT OUTER, Relation2_name
BY customer_id;
```

对 customers 和 orders 两个关系执行左外连接操作，代码如下。

```
grunt> outer_left = JOIN customers BY id LEFT OUTER, orders BY customer_id;
```

执行“DUMP outer_left;”命令，输出对应关系的信息，如下所示。

```
(1,Ramesh,32,Ahmedabad,2000,,,,)
(2,Khilan,25,Delhi,1500,101,2009-11-20 00:00:00,2,1560)
(3,Kaushik,23,Kota,2000,100,2009-10-08 00:00:00,3,1500)
(3,Kaushik,23,Kota,2000,102,2009-10-08 00:00:00,3,3000)
(4,Chaitali,25,Mumbai,6500,103,2008-05-20 00:00:00,4,2060)
(5,Hardik,27,Bhopal,8500,,,,)
(6,Komal,22,MP,4500,,,,)
```

```
(7,Muffy,24,Indore,10000,,,,)
```

ⓑ 右外连接：其将返回右表中的所有行，即使左表中没有匹配项。其语法格式如下。

```
grunt> outer_right = JOIN customers BY id RIGHT OUTER, orders BY customer_id;
```

对 customers 和 orders 执行右外连接操作，代码如下。

```
grunt> outer_right = JOIN customers BY id RIGHT OUTER, orders BY customer_id;
```

执行“DUMP outer_right;”命令，输出如下对应关系的信息。

```
(2,Khilan,25,Delhi,1500,101,2009-11-20 00:00:00,2,1560)
(3,Kaushik,23,Kota,2000,100,2009-10-08 00:00:00,3,1500)
(3,Kaushik,23,Kota,2000,102,2009-10-08 00:00:00,3,3000)
(4,Chaitali,25,Mumbai,6500,103,2008-05-20 00:00:00,4,2060)
```

ⓒ 全外连接：当一个关系中存在匹配时，全外连接操作将返回所有行。其语法格式如下。

```
grunt> outer_full = JOIN customers BY id FULL OUTER, orders BY customer_id;
```

对 customers 和 orders 执行全外连接操作，代码如下。

```
grunt> outer_full = JOIN customers BY id FULL OUTER, orders BY customer_id;
```

执行“DUMP outer_full;”命令，输出如下对应关系的信息。

```
(1,Ramesh,32,Ahmedabad,2000,,,,)
(2,Khilan,25,Delhi,1500,101,2009-11-20 00:00:00,2,1560)
(3,Kaushik,23,Kota,2000,100,2009-10-08 00:00:00,3,1500)
(3,Kaushik,23,Kota,2000,102,2009-10-08 00:00:00,3,3000)
(4,Chaitali,25,Mumbai,6500,103,2008-05-20 00:00:00,4,2060)
(5,Hardik,27,Bhopal,8500,,,,)
(6,Komal,22,MP,4500,,,,)
(7,Muffy,24,Indore,10000,,,,)
```

② GROUP 运算符。

GROUP 运算符用于在一个或多个关系中对数据进行分组，它收集具有相同 key 的数据。其语法格式如下。

```
grunt> Group_data = GROUP Relation_name BY age;
```

在 HDFS 目录/pig_data/中有一个名称为 student_details.txt 的文件，其内容如下所示。

```
001,Rajiv,Reddy,21,9848022337,Hyderabad
```

```
002,Siddarth,Battacharya,22,9848022338,Kolkata
003,Rajesh,Khanna,22,9848022339,Delhi
004,Preethi,Agarwal,21,9848022330,Pune
005,Trupthi,Mohanthy,23,9848022336,Bhuwaneshwar
006,Archana,Mishra,23,9848022335,Chennai
007,Komal,Nayak,24,9848022334,Trivendram
008,Bharathi,Nambiayar,24,9848022333,Chennai
```

将这个文件加载到 Apache Pig 中，关系名称为 student_details，代码如下。

```
grunt> student_details = LOAD 'hdfs://localhost:9000/pig_data/student_details.txt'
USING PigStorage(',') as (id:int, firstname:chararray, lastname:chararray, age:int,
phone:chararray, city:chararray);
```

按照年龄关系中的记录/元组进行分组，代码如下。

```
grunt> group_data = GROUP student_details by age;
```

执行“DUMP group_data;”命令，输出如下对应关系的信息。

```
(21,{(4,Preethi,Agarwal,21,9848022330,Pune),(1,Rajiv,Reddy,21,9848022337,Hydera
bad)})
(22,{(3,Rajesh,Khanna,22,9848022339,Delhi),(2,Siddarth,Battacharya,22,984802233
8,Kolkata)})
(23,{(6,Archana,Mishra,23,9848022335,Chennai),(5,Trupthi,Mohanthy,23,9848022336,
Bhuwaneshwar)})
(24,{(8,Bharathi,Nambiayar,24,9848022333,Chennai),(7,Komal,Nayak,24,9848022334,
Trivendram)})
```

可以观察到结果模式有两列：一列是 age，通过它对关系进行分组；另一列是 bag，包含一组元组，有各自年龄的学生记录。

使用“Describe”命令查看分组关系，可以看到表的模式，代码如下。

```
grunt> Describe group_data;
group_data: {group: int,student_details: {(id: int,firstname: chararray,
        lastname: chararray,age: int,phone: chararray,city: chararray)}}
```

按年龄和城市对关系进行多列分组，代码如下。

```
grunt> group_multiple = GROUP student_details by (age, city);
```

执行“DUMP group_multiple;”命令，输出对应关系的信息，如下所示。

```
grunt> DUMP group_multiple;
```

```
((21,Pune),{(4,Preethi,Agarwal,21,9848022330,Pune)})
((21,Hyderabad),{(1,Rajiv,Reddy,21,9848022337,Hyderabad)})
((22,Delhi),{(3,Rajesh,Khanna,22,9848022339,Delhi)})
((22,Kolkata),{(2,Siddarth,Battacharya,22,9848022338,Kolkata)})
((23,Chennai),{(6,Archana,Mishra,23,9848022335,Chennai)})
((23,Bhuwaneshwar),{(5,Trupthi,Mohanthy,23,9848022336,Bhuwaneshwar)})
((24,Chennai),{(8,Bharathi,Nambiayar,24,9848022333,Chennai)})
((24,Trivendram),{(7,Komal,Nayak,24,9848022334,Trivendram)})
```

按所有的列对关系进行分组，代码如下。

```
grunt> group_all = GROUP student_details All;
```

执行“DUMP group_all;”命令，输出如下对应关系的信息。

```
grunt> DUMP group_all;
(all,{(8,Bharathi,Nambiayar,24,9848022333,Chennai),(7,Komal,Nayak,24,9848022334 ,
Trivendram),
(6,Archana,Mishra,23,9848022335,Chennai),(5,Trupthi,Mohanthy,23,9848022336,Bhu
waneshwar),
(4,Preethi,Agarwal,21,9848022330,Pune),(3,Rajesh,Khanna,22,9848022339,Delhi),
(2,siddarth,Battacharya,22,9848022338,Kolkata),(1,Rajiv,Reddy,21,9848022337,Hyder
abad)})
```

③ COGROUP 运算符。

COGROUP 运算符的运作方式与 GROUP 运算符相同。这两个运算符之间的唯一区别是 GROUP 运算符通常用于涉及一个关系的语句，而 COGROUP 运算符用于涉及两个或多个关系的语句。

使用 COGROUP 分组两个关系，在 HDFS 目录/pig_data/中有两个文件，即 student_details.txt 和 employee_details.txt，其中 student_details.txt 文件内容如下所示。

```
001,Rajiv,Reddy,21,9848022337,Hyderabad
002,Siddarth,Battacharya,22,9848022338,Kolkata
003,Rajesh,Khanna,22,9848022339,Delhi
004,Preethi,Agarwal,21,9848022330,Pune
005,Trupthi,Mohanthy,23,9848022336,Bhuwaneshwar
```

```
006,Archana,Mishra,23,9848022335,Chennai
007,Komal,Nayak,24,9848022334,Trivendram
008,Bharathi,Nambiayar,24,9848022333,Chennai
```

employee_details.txt 文件内容如下所示。

```
001,Robin,22,newyork
002,BOB,23,Kolkata
003,Maya,23,Tokyo
004,Sara,25,London
005,David,23,Bhuwaneshwar
006,Maggy,22,Chennai
```

将这些文件分别加载到Pig中，关系名称分别为student_details和employee_details，代码如下。

```
grunt> student_details = LOAD 'hdfs://localhost:9000/pig_data/student_details.txt'
USING PigStorage(',') as (id:int, firstname:chararray, lastname:chararray, age:int,
phone:chararray, city:chararray);

grunt> employee_details = LOAD 'hdfs://localhost:9000/pig_data/employee_details.txt'
USING PigStorage(',') as (id:int, name:chararray, age:int, city:chararray);
```

将 student_details 和 employee_details 关系的记录/元组按关键字 age 进行分组，代码如下。

```
grunt> cogroup_data = COGROUP student_details by age, employee_details by age;
```

执行“DUMP cogroup_data;”命令，输出如下对应关系的信息。

```
(21,{(4,Preethi,Agarwal,21,9848022330,Pune),
(1,Rajiv,Reddy,21,9848022337,Hyderabad)},
    {    })
(22,{ (3,Rajesh,Khanna,22,9848022339,Delhi), (2,Siddarth,Battacharya,22,
9848022338,Kolkata) },
    { (6,Maggy,22,Chennai),(1,Robin,22,newyork) })
(23,{(6,Archana,Mishra,23,9848022335,Chennai),(5,Trupthi,Mohanthy,23,9848022336,
Bhuwaneshwar)},
    {(5,David,23,Bhuwaneshwar),(3,Maya,23,Tokyo),(2,BOB,23,Kolkata)})
(24,{(8,Bharathi,Nambiayar,24,9848022333,Chennai),(7,Komal,Nayak,24,9848022334,
```

```
Trivendram)},
        { })
    (25,{    },
        {(4,Sara,25,London)})
```

COGROUP 运算符根据年龄对来对每个关系的元组进行分组，其中每个组都描述了特定的年龄值。

考虑结果的第一个元组，其按照年龄 21 进行分组，包含两个包：第一个包保存了具有年龄为 21 岁的第一关系（即 student_details）的所有元组；第二个包保存了具有年龄为 21 岁的第二关系（即 employee_details）的所有元组。如果关系不具有年龄值为 21 的元组，则返回一个空包。

（4）排序

① ORDER BY 运算符。

ORDER BY 运算符用于基于一个或多个字段排序显示关系的内容。其语法格式如下。

```
grunt> Relation_name2 = ORDER Relation_name1 BY (ASC|DESC);
```

在 HDFS 目录/pig_data/中有一个名称为 student_details.txt 的文件，其内容如下所示。

```
001,Rajiv,Reddy,21,9848022337,Hyderabad
002,Siddarth,Battacharya,22,9848022338,Kolkata
003,Rajesh,Khanna,22,9848022339,Delhi
004,Preethi,Agarwal,21,9848022330,Pune
005,Trupthi,Mohanthy,23,9848022336,Bhuwaneshwar
006,Archana,Mishra,23,9848022335,Chennai
007,Komal,Nayak,24,9848022334,Trivendram
008,Bharathi,Nambiayar,24,9848022333,Chennai
```

通过关系 student_details 将此文件加载到 Pig 中，代码如下。

```
grunt> student_details = LOAD 'hdfs://localhost:9000/pig_data/student_details.txt' USING PigStorage(',') as (id:int, firstname:chararray, lastname:chararray, age:int, phone:chararray, city:chararray);
```

根据学生的年龄按降序排列关系，并使用 ORDER BY 运算符将关系存储到名为 order_by_data 的关系中，代码如下。

```
grunt> order_by_data = ORDER student_details BY age DESC;
```

执行“DUMP order_by_data;”命令，输出如下对应关系的信息。

```
(8,Bharathi,Nambiayar,24,9848022333,Chennai)
(7,Komal,Nayak,24,9848022334,Trivendram)
(6,Archana,Mishra,23,9848022335,Chennai)
(5,Trupthi,Mohanthy,23,9848022336,Bhuwaneshwar)
(3,Rajesh,Khanna,22,9848022339,Delhi)
(2,Siddarth,Battacharya,22,9848022338,Kolkata)
(4,Preethi,Agarwal,21,9848022330,Pune)
(1,Rajiv,Reddy,21,9848022337,Hyderabad)
```

② LIMIT 运算符。

LIMIT 运算符用于从关系中获取有限数量的元组。其语法格式如下。

```
grunt> Result = LIMIT Relation_name required number of tuples;
```

在 HDFS 目录/pig_data/中有一个名称为 student_details.txt 的文件，通过关系 student_details 将此文件加载到 Pig 中，代码如下。

```
grunt> student_details = LOAD 'hdfs://localhost:9000/pig_data/student_details.txt'
USING PigStorage(',') as (id:int, firstname:chararray, lastname:chararray,age:int,
phone:chararray, city:chararray);
```

使用 LIMIT 运算符将从关系中获取数量为 4 的元组存储到 limit_data 关系中，代码如下。

```
grunt> limit_data = LIMIT student_details 4;
```

执行“DUMP limit_data;”命令，输出对应关系的信息，如下所示。

```
(1,Rajiv,Reddy,21,9848022337,Hyderabad)
(2,Siddarth,Battacharya,22,9848022338,Kolkata)
(3,Rajesh,Khanna,22,9848022339,Delhi)
(4,Preethi,Agarwal,21,9848022330,Pune)
```

（5）组合和分隔

① UNION 运算符。

Pig 中的 UNION 运算符用于合并两个关系的内容。要对两个关系执行 UNION 操作，它们的列和域必须相同。其语法格式如下。

```
grunt> Relation_name3 = UNION Relation_name1, Relation_name2;
```

HDFS 的/pig_data/目录中有两个文件，即 student_data1.txt 和 student_data2.txt，其中 student_data1.txt 文件内容如下所示。

```
001,Rajiv,Reddy,9848022337,Hyderabad
002,Siddarth,Battacharya,9848022338,Kolkata
003,Rajesh,Khanna,9848022339,Delhi
004,Preethi,Agarwal,9848022330,Pune
005,Trupthi,Mohanthy,9848022336,Bhuwaneshwar
006,Archana,Mishra,9848022335,Chennai
```

student_data2.txt 文件内容如下所示。

```
7,Komal,Nayak,9848022334,Trivendram
8,Bharathi,Nambiayar,9848022333,Chennai
```

通过关系 student1 和 student2 将这两个文件加载到 Pig 中，代码如下。

```
grunt> student1 = LOAD 'hdfs://localhost:9000/pig_data/student_data1.txt' USING PigStorage(',') as (id:int, firstname:chararray, lastname:chararray, phone:chararray, city:chararray);

grunt> student2 = LOAD 'hdfs://localhost:9000/pig_data/student_data2.txt' USING PigStorage (',') as (id:int, firstname:chararray, lastname:chararray, phone:chararray, city:chararray);
```

使用 UNION 运算符合并这两个关系的内容，代码如下。

```
grunt> student =UNION student1, student2;
```

执行“DUMP student;”命令，输出对应关系的信息，如下所示。

```
(1,Rajiv,Reddy,9848022337,Hyderabad)
(2,Siddarth,Battacharya,9848022338,Kolkata)
(3,Rajesh,Khanna,9848022339,Delhi)
(4,Preethi,Agarwal,9848022330,Pune)
(5,Trupthi,Mohanthy,9848022336,Bhuwaneshwar)
(6,Archana,Mishra,9848022335,Chennai)
(7,Komal,Nayak,9848022334,Trivendram)
(8,Bharathi,Nambiayar,9848022333,Chennai)
```

② SPLIT 运算符。

SPLIT 运算符用于将关系拆分为两个或多个关系。其语法格式如下。

```
grunt> SPLIT Relation1_name INTO Relation2_name IF (condition1), Relation2_name (condition2);
```

在 HDFS 目录/pig_data/中有一个名称为 student_details.txt 的文件，通过关

系 student_details 将此文件加载到 Pig 中，代码如下。

```
student_details = LOAD 'hdfs://localhost:9000/pig_data/student_details.txt' USING PigStorage(',') as (id:int, firstname:chararray, lastname:chararray, age:int, phone:chararray, city:chararray);
```

将关系分为两个：一个列出年龄小于 23 岁的员工，另一个列出年龄为 22～25 岁的员工。代码如下。

```
SPLIT student_details into student_details1 if age<23, student_details2 if (22<age and age<25);
```

执行“DUMP student_details1;”命令，输出对应关系的信息，如下所示。

```
grunt> DUMP student_details1;
(1,Rajiv,Reddy,21,9848022337,Hyderabad)
(2,Siddarth,Battacharya,22,9848022338,Kolkata)
(3,Rajesh,Khanna,22,9848022339,Delhi)
(4,Preethi,Agarwal,21,9848022330,Pune)
```

执行“DUMP student_details2;”命令，输出对应关系的信息，如下所示。

```
grunt> DUMP student_details2;
(5,Trupthi,Mohanthy,23,9848022336,Bhuwaneshwar)
(6,Archana,Mishra,23,9848022335,Chennai)
(7,Komal,Nayak,24,9848022334,Trivendram)
(8,Bharathi,Nambiayar,24,9848022333,Chennai)
```

任务实施

1. 环境配置及数据准备

在/usr/local/Hadoop/etc/Hadoop/目录中，编辑 mapred-site.xml 文件，代码如下。

```
<property>
    <name>mapred.job.tracker</name>
    <value>127.0.0.1:9001</value>
<property>
```

执行“start-all.sh”命令，启动 Hadoop，再执行“mr-jobhistory-daemon.sh start historyserver”命令，启动 HistoryServer 服务，如图 4-1 所示（start-all.sh

命令不会启动 HistoryServer 服务）。

执行“jps”命令，查看 Hadoop 系统进程，如图 4-2 所示。

```
:/usr/local$ mr-jobhistory-daemon.sh start historyserver
starting historyserver, logging to /usr/local/hadoop/hadoop-2.7.1/logs/mapred-a-
historyserver-a.out
:/usr/local$
```

图 4-1　启动 HistoryServer 服务

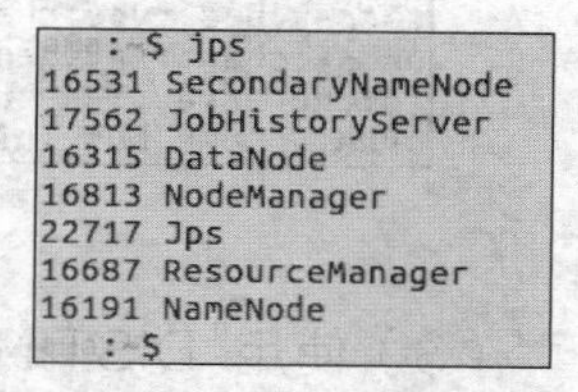

图 4-2　查看系统进程

执行“hadoop dfs -mkdir /pig”命令，在 HDFS 中创建 pig 目录，并执行“-put”命令，将本地数据上传到 HDFS 的 pig 目录中，如图 4-3 所示。

```
:/usr/local$ hadoop dfs -mkdir /pig
DEPRECATED: Use of this script to execute hdfs command is deprecated.
Instead use the hdfs command for it.

:/usr/local$ hadoop dfs -put /home/a/bus_info.csv /pig
DEPRECATED: Use of this script to execute hdfs command is deprecated.
Instead use the hdfs command for it.

:/usr/local$ hadoop dfs -ls /pig
DEPRECATED: Use of this script to execute hdfs command is deprecated.
Instead use the hdfs command for it.

Found 1 items
-rw-r--r--   1 a supergroup     855557 2018-07-30 10:23 /pig/bus_info.csv
```

图 4-3　将本地数据上传到 HDFS 的 pig 目录中

2. 在 Pig 中预处理数据

（1）执行“pig -x mapreduce”命令，进入 Pig 的 MapReduce 执行模式（Pig 的另一种执行模式为 Local 模式），如图 4-4 所示。

```
:/usr/local$ pig -x mapreduce
18/07/30 10:20:02 INFO pig.ExecTypeProvider: Trying ExecType : LOCAL
18/07/30 10:20:02 INFO pig.ExecTypeProvider: Trying ExecType : MAPREDUCE
18/07/30 10:20:02 INFO pig.ExecTypeProvider: Picked MAPREDUCE as the ExecType
18/07/30 10:20:02 WARN pig.Main: Cannot write to log file: /usr/local/pig_153291
7202308.log
2018-07-30 10:20:02,310 [main] INFO  org.apache.pig.Main - Apache Pig version 0.
17.0 (r1797386) compiled Jun 02 2017, 15:41:58
2018-07-30 10:20:02,383 [main] INFO  org.apache.pig.impl.util.Utils - Default bo
otup file /home/a/.pigbootup not found
2018-07-30 10:20:03,642 [main] INFO  org.apache.hadoop.conf.Configuration.deprec
ation - mapred.job.tracker is deprecated. Instead, use mapreduce.jobtracker.addr
ess
2018-07-30 10:20:03,645 [main] INFO  org.apache.pig.backend.hadoop.executionengi
ne.HExecutionEngine - Connecting to hadoop file system at: hdfs://127.0.0.1:9000
2018-07-30 10:20:04,705 [main] INFO  org.apache.pig.PigServer - Pig Script ID fo
r the session: PIG-default-730b4f3b-d3d5-473c-806d-612200309ad3
2018-07-30 10:20:04,710 [main] WARN  org.apache.pig.PigServer - ATS is disabled
since yarn.timeline-service.enabled set to false
grunt>
```

图 4-4　Pig 的 MapReduce 执行模式

（2）将数据从 HDFS 中加载到 Pig 中，LOAD 为数据加载语句，USING PigStorage(',')表示数据是以“,”为分隔符的，chararray 为 Pig 的字符串或字符数组类型，关系创建完成后，可以执行“DUMP”命令验证关系，如图 4-5 所示。

```
grunt> bus_info = LOAD '/pig/bus_info.csv' USING PigStorage(',') as(bus_name:cha
rarray,bus_type:chararray,bus_time:chararray,tieck:chararray,licheng:chararray,g
ongsi:chararray,gengxin:chararray,wang_info:chararray,wang_buff:chararray,fan_in
fo:chararray,fan_buff:chararray);
grunt> DUMP bus_info;
```

图 4-5　加载数据及验证关系

（3）使用 DISTINCT 运算符从 bus_info 关系中删除冗余元组，将其另存在 distinct_data 关系中，并执行“DUMP”命令验证关系，如图 4-6 所示。

```
grunt> distinct_data = DISTINCT bus_info;
grunt> DUMP distinct_data;
```

图 4-6　删除冗余元组及验证关系

（4）使用 FILTER 运算符获取 distinct_data 关系中所有不为空的数据，将其另存在 filter_data 关系中，并执行“DUMP”命令验证关系，如图 4-7 所示。

```
grunt> filter_data = FILTER distinct_data BY bus_name != '' and bus_type != '' a
nd bus_time != '' and tieck != '' and licheng != '' and gongsi != '' and gengxin
 != '' and wang_info != '' and wang_buff != '' and fan_info != '' and fan_buff !
='';
grunt> DUMP filter_data;
```

图 4-7　FILTER 操作及验证关系

3. 将 Pig 预处理的文件导出到 HDFS 中

图 4-8 所示为将预处理的文件导出到 HDFS 中的命令。

```
grunt> STORE filter_data INTO 'hdfs://127.0.0.1:9000/pig_output' USING PigStorag
e(',');
```

图 4-8　将预处理的文件导出到 HDFS 中的命令

4. 查看数据

在 HDFS 中查看处理后的数据，并执行“cat”命令查看输出文件，图 4-9 所示为预处理后的数据。

```
:~$ hadoop dfs -cat /pig_output/part-r-00000
DEPRECATED: Use of this script to execute hdfs command is deprecated.
Instead use the hdfs command for it.

北京1路公交车路线,市区普线,运行时间: 老山公交场站5:00-23:00|四惠枢纽站5:00-23:0
0,票价信息: 10公里以内票价2元, 每增加5公里以内加价1元。持卡乘车普通卡5折、学生
卡2.5折优惠。最高票价6元。,北京公交集团第四客运分公司(服务热线:010-68683295),最
后更新: 2018-07-22,全程25.31公里。,北京1路公交车路线(老山公交场站-四惠枢纽站),
老山公交场站 老山南路东口 地铁八宝山站 玉泉路口西 永定路口东 五棵松桥西 沙沟路
口西 东翠路口 万寿路口西 翠微路口 公主坟 军事博物馆 木樨地西 工会大楼 南礼士路
复兴门内 西单路口东 天安门西 天安门东 东单路口西 北京站口东 日坛路 永安里路口西
 大北窑西 大北窑东 郎家园 四惠枢纽站 ,北京1路公交车路线(四惠枢纽站-老山公交场站
),四惠枢纽站 八王坟西 郎家园 大北窑东 大北窑西 永安里路口西 日坛路 北京站口东
东单路口西 天安门东 天安门西 西单路口东 复兴门内 南礼士路 工会大楼 木樨地西 军
事博物馆 公主坟 翠微路口 万寿路口西 东翠路口 沙沟路口西 五棵松桥东 永定路口东
玉泉路口西 地铁八宝山站 老山南路东口 老山公交场站
```

图 4-9　预处理后的数据

任务 2 用 Kettle 进行数据预处理

任务描述

（1）学习 Kettle 的相关基础知识。

（2）使用 Kettle 实现“北京公交线路信息”数据的预处理。

任务目标

（1）熟悉 Kettle 的相关基础知识。

（2）学会使用 Kettle 完成“北京公交线路信息”数据的预处理。

知识准备

Kettle 的中文名称为水壶。在数据处理中，Kettle 是一个 ETL 工具集，它可以管理来自不同数据库的数据，通过提供一个图形化的用户环境来描述想做什么，而不是怎么做。Kettle 中有两种脚本文件：Transformation 和 Job。其中，Transformation 用于完成针对数据的基础转换，而 Job 用于完成整个工作流的控制。

1. Kettle 的三大模块

（1）Spoon——转换/工作设计工具（GUI 方式）。

（2）Kitchen——工作执行器（命令行方式）。

（3）Pan——转换执行器（命令行方式）。

2. Kettle 的组件

创建一个新的 Transformation，Kettle 中 Transformation 文件的默认后缀名为 ktr。

创建一个新的 Job，Kettle 中 Job 文件的默认后缀名为 kjb。

（1）Transformation 组件树

① Main Tree：列出了一个 Transformation 的基本属性，可以通过各个节点进行查看。

② DB 连接：显示当前 Transformation 中的数据库连接，每一个 Transformation 的数据库连接都需要单独配置。

③ Steps：一个 Transformation 中应用到的环节列表。

④ Hops：一个 Transformation 中应用到的节点连接列表。

核心对象菜单列出的是 Transformation 中可以调用的环节列表，可以通过鼠标拖动的方式对环节进行添加。Transformation 的常用环节如表 4-2 所示。

表 4-2 Transformation 的常用环节

类别	环节名称	功能说明
Input	文本文件输入	从本地文本文件输入数据
	表输入	从数据库表中输入数据
	获取系统信息	读取系统信息并输入数据
Output	文本文件输出	将处理结果输出到文本文件中
	表输出	将处理结果输出到数据库表中
	插入/更新	根据处理结果对数据库表进行插入/更新操作，如果数据库中不存在相关记录则插入，否则更新。可根据查询条件中的字段进行判断
	更新	根据处理结果对数据库进行更新，若需要更新的数据在数据库表中无记录，则会报错并终止
	删除	根据处理结果对数据库记录进行删除，若需要删除的数据在数据库表中无记录，则会报错并终止
Lookup	数据库查询	根据设定的查询条件，对目标表进行查询，返回需要的结果字段
	流查询	将目标表读取到内存中，通过查询条件对内存中的数据集进行查询
	调用 DB 存储过程	调用数据库存储过程
Transform	字段选择	选择需要的字段，过滤掉不需要的字段，也可做数据库字段对应
	过滤记录	根据条件对记录进行过滤
	排序记录	使数据根据某些条件进行排序
	空操作	无操作
	增加常量	增加需要的常量字段
Scripting	Modified JavaScript Value	扩展功能，编写 JavaScript，对数据进行相应处理
Mapping	映射（子转换）	数据映射
Job	Set Variables	设置环境变量
	Get Variables	获取环境变量

（2）Job 组件树

① Main Tree：列出了一个 Job 的基本属性，可以通过各个节点进行查看。

② DB 连接：显示当前 Job 中的数据库连接，每一个 Job 的数据库连接都需要单独配置。

③ Job Entries：即作业项目一个 Job 中引用的环节列表。

核心对象菜单列出的是 Job 中可以调用的环节列表，可以通过鼠标拖动的方式对环节进行添加。每一个环节都可以通过鼠标拖动操作被添加到主窗口中，并可通过“Shift”键+鼠标拖动，实现环节之间的连接。Job 中可以调用的常用环节如表 4-3 所示。

表 4-3　Job 中可以调用的常用环节

类别	环节名称	功能说明
作业项目	START	开始
	DUMMY	结束
	Transformation	引用 Transformation 流程
	Job	引用 Job 流程
	Shell	调用 Shell 脚本
	SQL	执行 SQL 语句
	FTP	通过 FTP 下载
	Table exists	检查目标表是否存在，返回布尔值
	File exists	检查文件是否存在，返回布尔值
	JavaScript	执行 JavaScript
	Create file	创建文件
	Delete file	删除文件
	Wait for file	等待文件，文件出现后继续下一个环节
	File compare	文件比较，返回布尔值
	Wait for	等待时间，设定一段时间，使 Kettle 流程处于等待状态
	Zip file	压缩文件为 ZIP 格式

任务实施

1. 修改数据库默认字符集并创建数据库

数据存入 MySQL 时会出现中文乱码问题，因此需要将 MySQL 数据库的默认字符集设为 UTF-8。具体操作如下。

（1）进入默认 MySQL 安装目录，查看 my.cnf 文件，如图 4-10 所示。

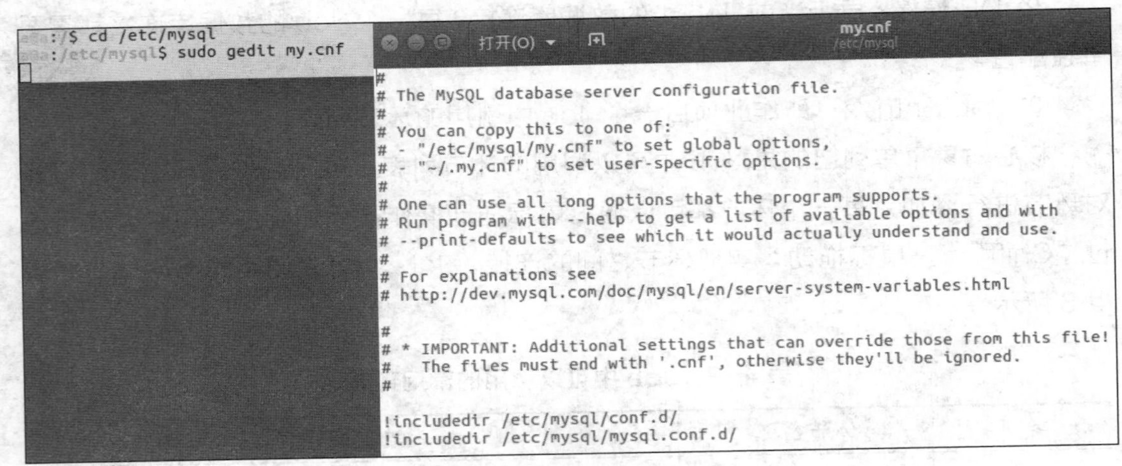

图 4-10　查看 my.cnf 文件

从图 4-10 可以看出 my.cnf 文件引用了 conf.d 与 mysql.conf.d 两个目录中的文件。

（2）编辑 conf.d 目录中的 mysql.cnf 文件，如图 4-11 所示。

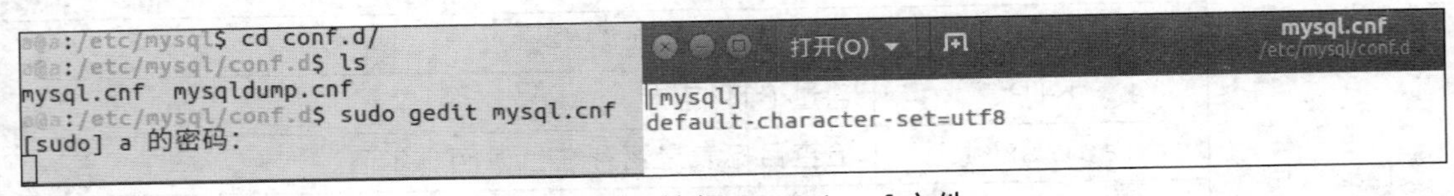

图 4-11　编辑 mysql.cnf 文件

（3）修改 mysql.conf.d 目录中的 mysqld.cnf 文件，在[mysqld]中添加“character-set-server=utf8”，如图 4-12 所示。

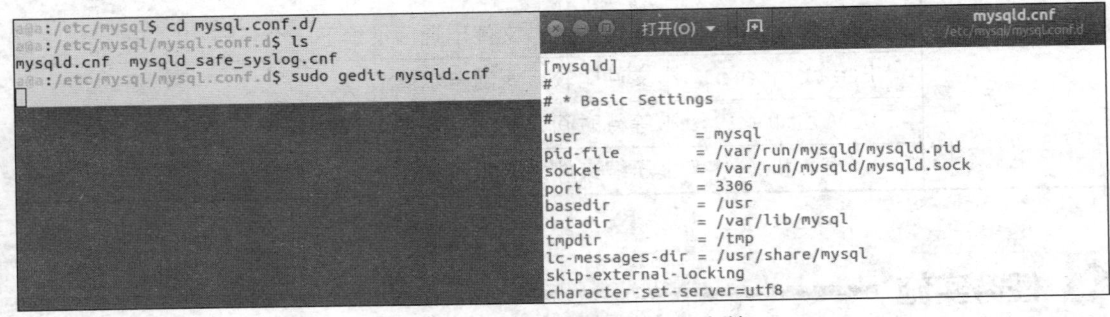

图 4-12　修改 mysqld.cnf 文件

（4）进入 MySQL 数据库，执行“show variables like'%char%';”命令，查看数据库字符集，如图 4-13 所示。

```
mysql> show variables like '%char%';
+--------------------------------------+----------------------------+
| Variable_name                        | Value                      |
+--------------------------------------+----------------------------+
| character_set_client                 | utf8                       |
| character_set_connection             | utf8                       |
| character_set_database               | utf8                       |
| character_set_filesystem             | binary                     |
| character_set_results                | utf8                       |
| character_set_server                 | utf8                       |
| character_set_system                 | utf8                       |
| character_sets_dir                   | /usr/share/mysql/charsets/ |
| validate_password_special_char_count | 1                          |
+--------------------------------------+----------------------------+
9 rows in set (0.08 sec)

mysql>
```

图 4-13　查看数据库字符集

2. 创建数据表

进入 student 数据库，创建 bus_info 数据表，用于存储执行数据清洗操作后的数据，如图 4-14 所示。

```
mysql> use student;
Database changed
mysql> create table bus_info (bus_name varchar(255),bus_type varchar(255),bus_ti
me varchar(255),ticket varchar(255),gongsi varchar(255),gengxin varchar(255),lic
heng varchar(1000),wang_info varchar(255),wang_buff varchar(1000),fan_info varch
ar(255),fan_buff varchar(1000)) DEFAULT CHARSET=utf8;
Query OK, 0 rows affected (0.09 sec)

mysql>
```

图 4-14　创建 bus_info 数据表

3. 创建 Transformations

（1）进入/usr/local/kettle 目录，执行“./spoon.sh”命令，启动 Kettle，如图 4-15 所示。

```
a@a:~$ cd /usr/local/kettle/
a@a:/usr/local/kettle$ ./spoon.sh
#######################################################################
WARNING:  no libwebkitgtk-1.0 detected, some features will be unavailable
    Consider installing the package with apt-get or yum.
    e.g. 'sudo apt-get install libwebkitgtk-1.0-0'
#######################################################################
Java HotSpot(TM) 64-Bit Server VM warning: ignoring option MaxPermSize=256m; sup
port was removed in 8.0
10:46:24,021 INFO  [KarafBoot] Checking to see if org.pentaho.clean.karaf.cache
is enabled
10:46:24,256 INFO  [KarafInstance]
*******************************************************************************
*** Karaf Instance Number: 1 at /usr/local/kettle/./system/karaf/caches/spo ***
***   on/data-1                                                             ***
*** Karaf Port:8802                                                         ***
*** OSGI Service Port:9051                                                  ***
*******************************************************************************
```

图 4-15　启动 Kettle

（2）新建 Transformations。

首先，新建一个 Transformations（双击 Transformations 图标即可）。

其次，选择“Design”选项卡，先后将“Input”节点中的“Text file input”、“Transform”节点中的“Unique rows”、“Utility”节点中的“If field value is null”及“Output”节点中的“Table output”拖动到工作区中。

最后，按住“Shift”键，拖动“Text file input”图标到“Unique rows”图标，进行连线，以此操作实现各个环节间的连线，如图 4-16 所示。

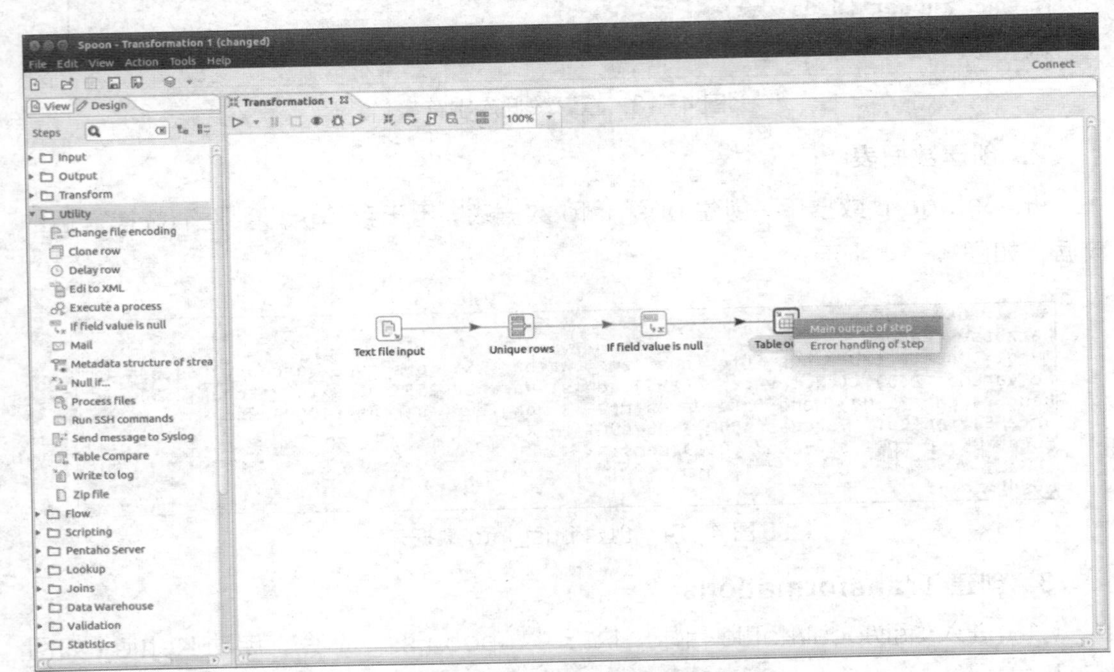

图 4-16　实现各个环节间的连线

（3）设置“Text file input”。

双击“Text file input”图标，进入“Text file input”设置界面，如图 4-17 所示。

在“Text file input”界面中单击“Browse...”按钮，设置目录为“公交线路数据”所在目录，并选中要清洗的资源文件“bus_info.csv”，单击“确定”按钮。在“Text file input”界面中单击“Add”按钮，导入资源文件，如图 4-18 所示。

选择“Content”选项卡，设置 Separator 为“,”（Separator 表示文件分隔符），设置 Format 为“mixed”，设置 Encoding 为“UTF-8”（这里设置的字符集编码要和文件字符集的编码一致），如图 4-19 所示，设置完毕后单击“OK”按钮。

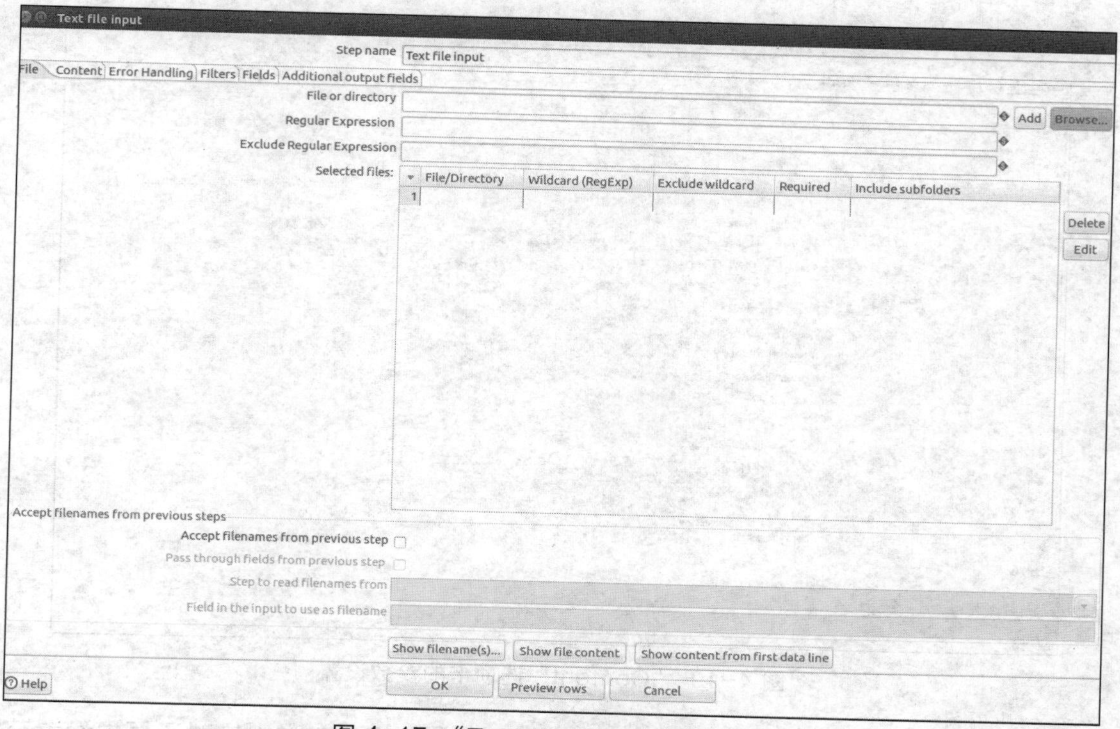

Text file input

Step name Text file input

File | Content | Error Handling | Filters | Fields | Additional output fields

File or directory

Regular Expression Add Browse...

Exclude Regular Expression

Selected files:

	File/Directory	Wildcard (RegExp)	Exclude wildcard	Required	Include subfolders
1					

Delete

Edit

Accept filenames from previous steps

Accept filenames from previous step

Pass through fields from previous step

Step to read filenames from

Field in the input to use as filename

Show filename(s)... Show file content Show content from first data line

Help OK Preview rows Cancel

图 4-17 “Text file input”设置界面

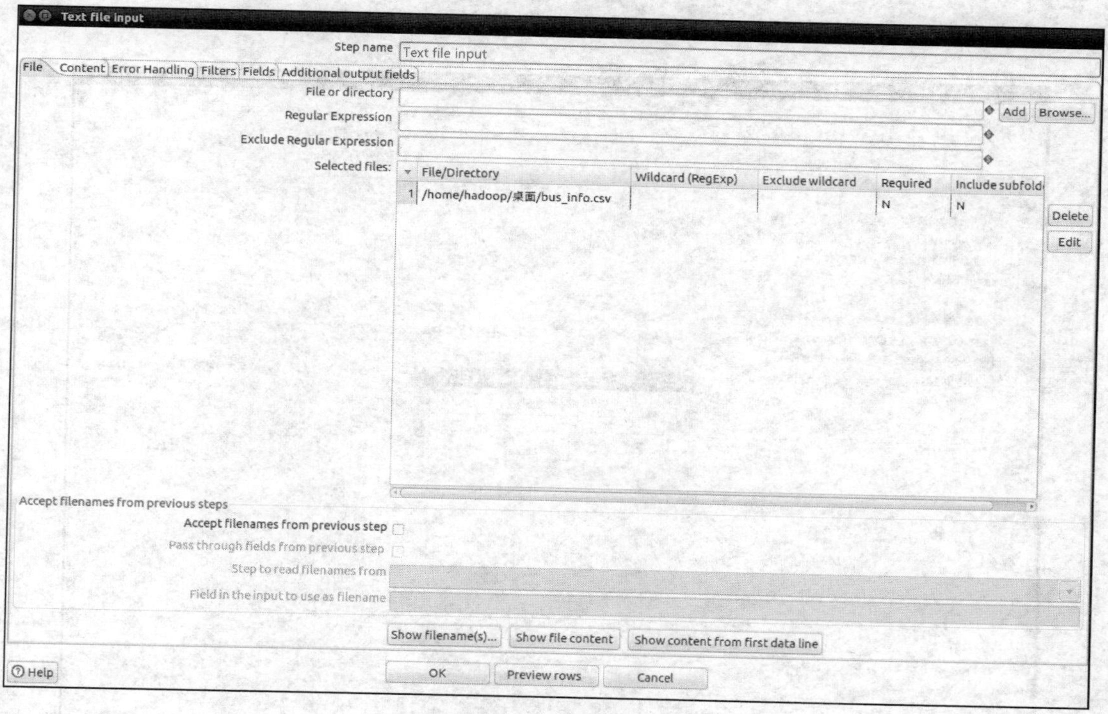

Text file input

Step name Text file input

File | Content | Error Handling | Filters | Fields | Additional output fields

File or directory

Regular Expression Add Browse...

Exclude Regular Expression

Selected files:

	File/Directory	Wildcard (RegExp)	Exclude wildcard	Required	Include subfold
1	/home/hadoop/桌面/bus_info.csv			N	N

Delete

Edit

Accept filenames from previous steps

Accept filenames from previous step

Pass through fields from previous step

Step to read filenames from

Field in the input to use as filename

Show filename(s)... Show file content Show content from first data line

Help OK Preview rows Cancel

图 4-18 导入资源文件

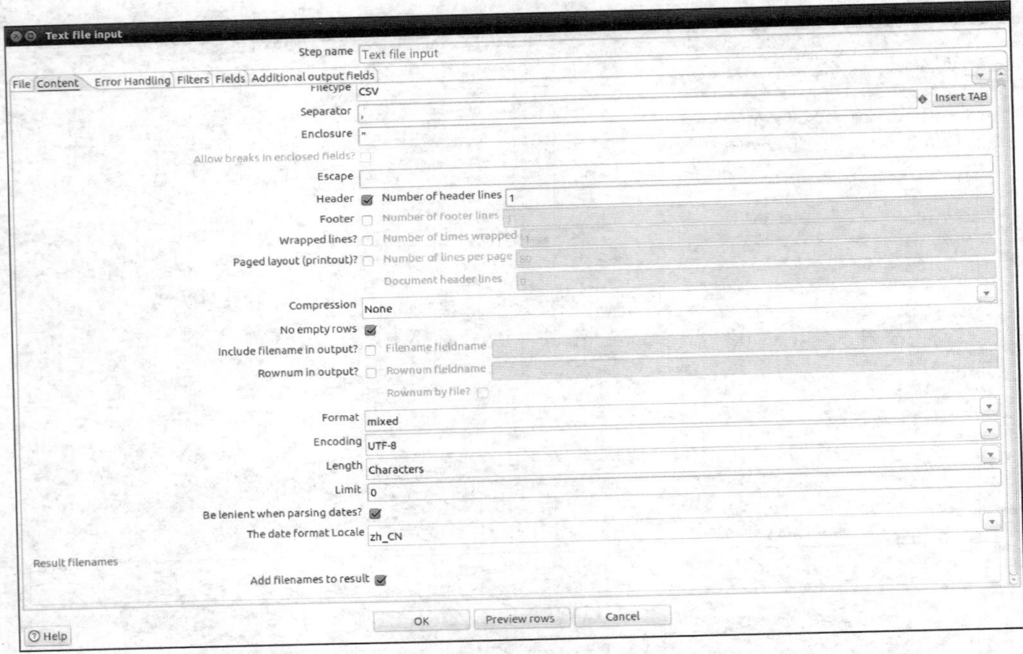

图 4-19 “Content”选项卡的参数设置

选择“Fields”选项卡，并单击“Get Fields”按钮，在弹出的对话框中设置获取数据的行数，这里设置为“100”，设置完毕后单击“OK”按钮，如图 4-20 所示。

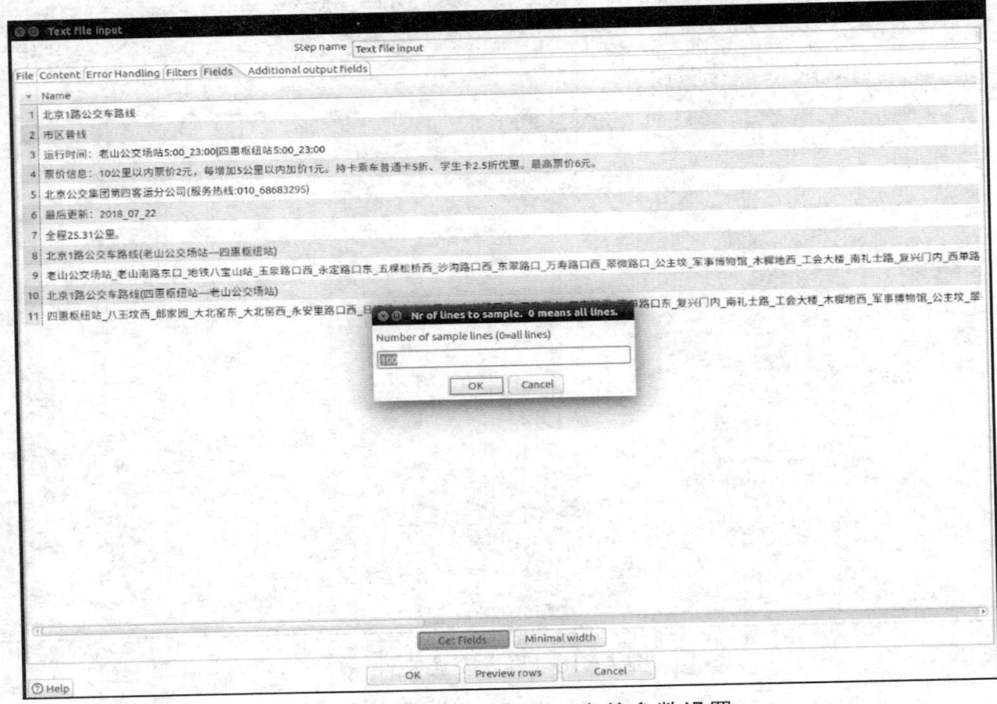

图 4-20 “Fields”选项卡的参数设置

再次选择“Content”选项卡，单击“Preview rows”按钮，预览数据，查看完毕后单击“Close”按钮，随后单击“OK”按钮，如图 4-21 所示。

图 4-21　预览数据

（4）设置“Unique rows”的相关参数。

双击“Unique rows”图标，进入“Unique rows”设置界面，Unique rows 组件的功能是去重，单击“Get”按钮，获得字段名，设置完毕后单击“OK”按钮，如图 4-22 所示。

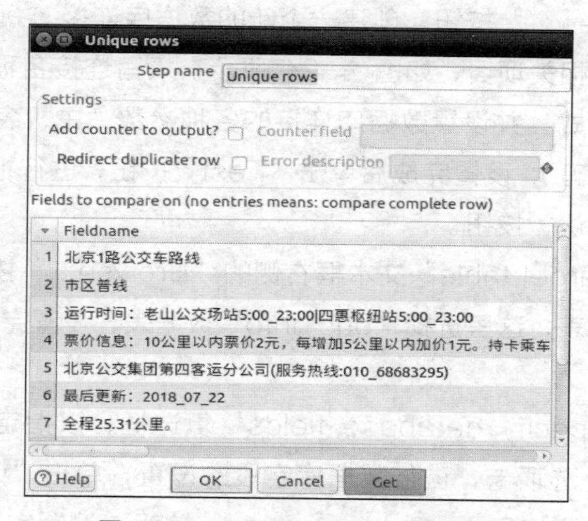

图 4-22　Unique rows 设置界面

（5）双击“If field value is null”图标，进入“Replace null value”设置界面，设置“If field value is null”相关内容，将“Replace by value”设置为“空值”（即将数据中所有的空数据填充为“空值”），如图 4-23 所示。

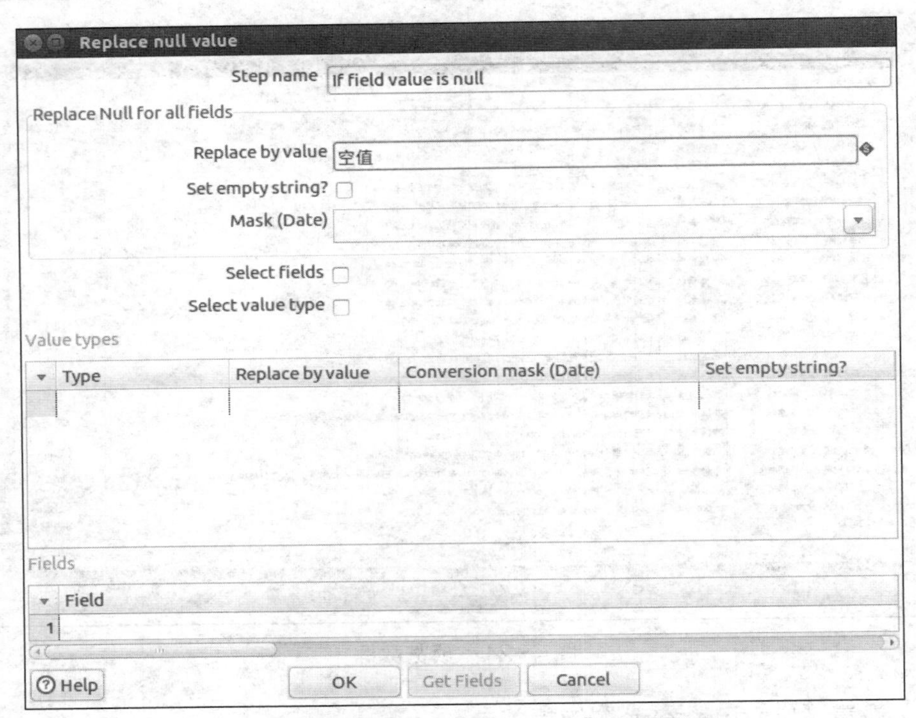

图 4-23 “Replace null value”设置界面

（6）双击“Table output”图标，进入“Table output”设置界面，设置“Table output”相关参数。

首先，单击“New...”按钮，创建一个新的数据库连接。

其次，依次设置相关项目，如图 4-24 所示。填写连接名称，选择存储数据的数据库，选择连接方式，并设置数据库连接的各种参数（主机名称、数据库名称、端口号、用户名和密码），设置完成后单击“Test”按钮，进行相关设置的测试。测试成功后，单击“确认”按钮。

再次，单击“Target table”文本框右侧的“Browse...”按钮，选择数据清洗操作后要保存的数据表，这里选择“bus_info”数据表，设置完成后单击“OK”按钮，如图 4-25 所示。

最后，选中“Specify database fields”复选框，以指定数据库字段，选择“Database fields”选项卡，进行数据库字段的设置，单击“Get fields”按钮，获取数据库字段，单击“Enter field mapping”按钮，设置字段之间的映射关系，其中，“Source fields”为数据源中的相关信息，“Target fields”为数据库中的目标字段，设置好映射关系后，单击“Add”按钮，将对应关系添加到“Mappings”列表框中。按照此操作，设置好所有的映射关系，设置完成后单击“OK”按钮，如图 4-26 所示。

图 4-24　设置“Table output”相关参数

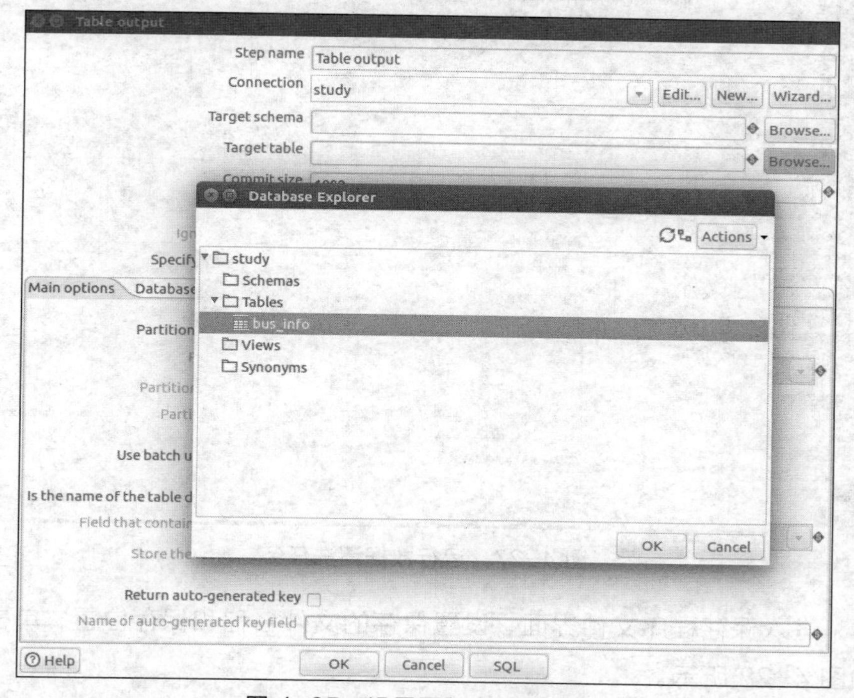

图 4-25　设置要保存的数据表

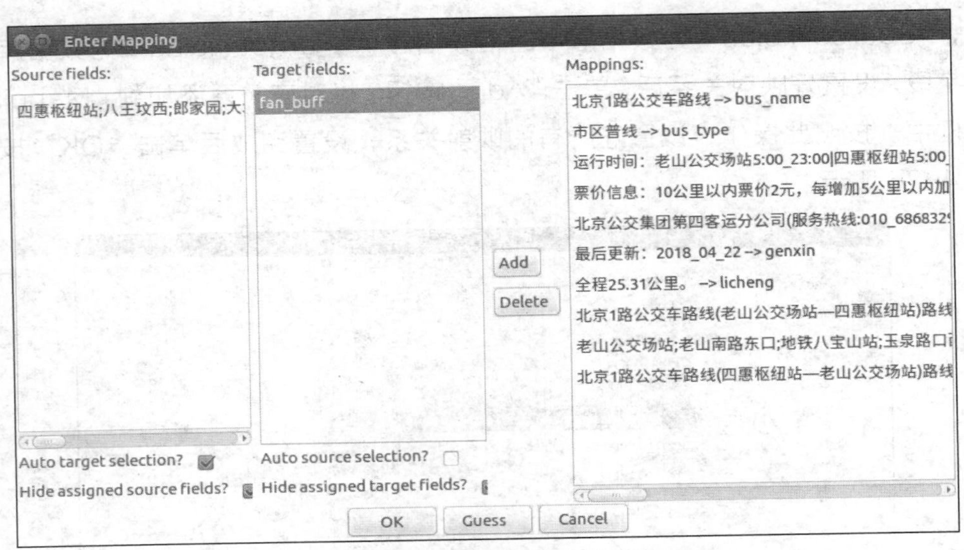

图 4-26　Database fields 的相关设置

4. 运行任务

单击窗口左上方的“▷”按钮，弹出“Run Options”对话框，单击“Run”按钮，运行数据清洗任务，如图 4-27 所示。

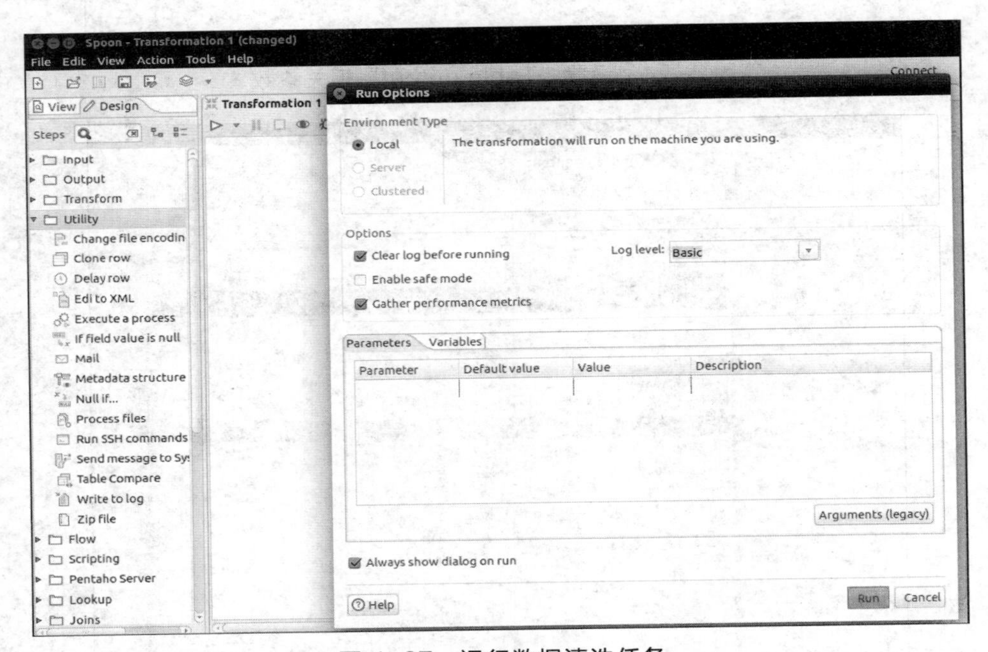

图 4-27　运行数据清洗任务

此时，进入保存程序文件界面，设置保存的文件名称和保存位置，单击“确定”按钮，如图 4-28 所示。

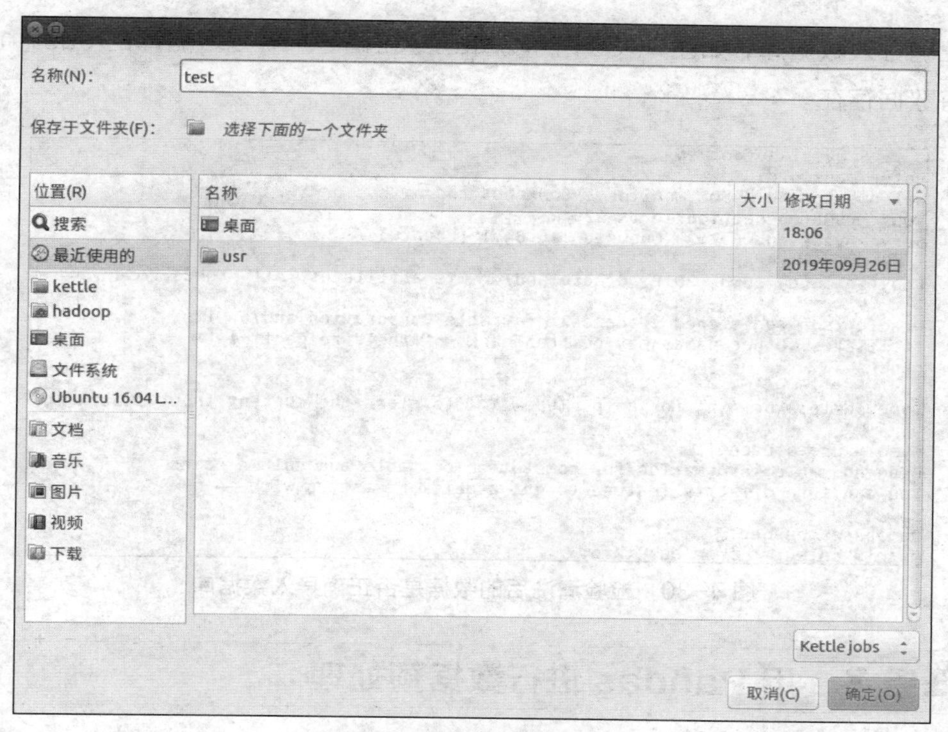

图 4-28　保存程序文件界面

执行数据清洗操作，并将清洗后的数据批量导入到数据库中，如图 4-29 所示。

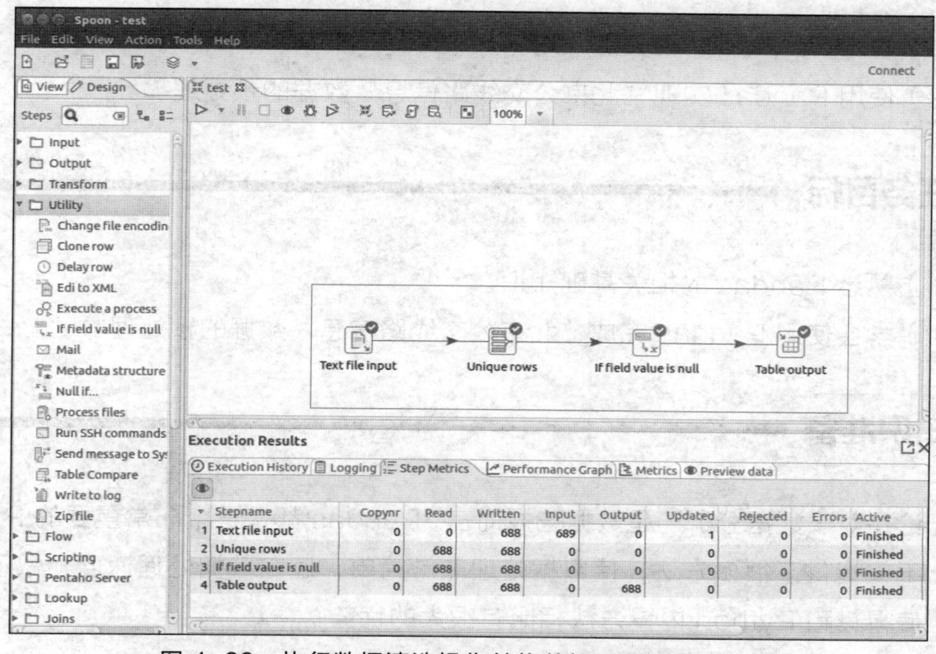

	Stepname	Copynr	Read	Written	Input	Output	Updated	Rejected	Errors	Active
1	Text file input	0	0	688	689	0	1	0	0	Finished
2	Unique rows	0	688	688	0	0	0	0	0	Finished
3	If field value is null	0	688	688	0	0	0	0	0	Finished
4	Table output	0	688	688	0	688	0	0	0	Finished

图 4-29　执行数据清洗操作并将数据导入到数据库中

重新打开一个终端，进入 MySQL，进入 student 数据库，检验清洗后的数据是否正确导入数据库，如图 4-30 所示。如已经正确导入，则数据库的 bus_info 表中会有数据存在。

```
  :/$ mysql -uroot -p
Enter password:
Welcome to the MySQL monitor.  Commands end with ; or \g.
Your MySQL connection id is 22
Server version: 5.7.23-0ubuntu0.16.04.1 (Ubuntu)

Copyright (c) 2000, 2018, Oracle and/or its affiliates. All rights reserved.

Oracle is a registered trademark of Oracle Corporation and/or its
affiliates. Other names may be trademarks of their respective
owners.

Type 'help;' or '\h' for help. Type '\c' to clear the current input statement.

mysql> use student;
Reading table information for completion of table and column names
You can turn off this feature to get a quicker startup with -A

Database changed
mysql> select * from bus_info;
```

图 4-30 检验清洗后的数据是否正确导入数据库

任务 3 用 Pandas 进行数据预处理

任务描述

（1）学习 Pandas 的相关基础知识。

（2）使用 Pandas 实现“北京公交线路信息”数据的预处理。

任务目标

（1）熟悉 Pandas 的相关基础知识。

（2）学会使用 Pandas 完成“北京公交线路信息”数据的预处理。

知识准备

无论是对于数据分析还是数据挖掘而言，Pandas 都是一个非常重要的 Python 包。它不仅提供了很多方法，使数据处理非常简单，还在数据处理速度上做了很多优化，使得其和 Python 内置方法相比有很大的优势。

1. 导入数据

pd.read_csv(filename)：从 CSV 文件中导入数据。

pd.read_table(filename)：从限定分隔符的文本文件中导入数据。

pd.read_excel(filename)：从 Excel 文件中导入数据。

pd.read_sql(query, connection_object)：从 SQL 表/库中导入数据。

pd.read_json(json_string)：从 JSON 格式的字符串中导入数据。

pd.read_html(url)：解析 URL、字符串或 HTML 文件，抽取其中的 tables。

pd.read_clipboard()：从粘贴板中获取内容，并传送给 read_table()。

pd.DataFrame(dict)：从字典对象中导入数据，Key 表示列名，Value 表示数据。

2. 导出数据

df.to_csv(filename)：导出数据到 CSV 文件中。

df.to_excel(filename)：导出数据到 Excel 文件中。

df.to_sql(table_name, connection_object)：导出数据到 SQL 表中。

df.to_json(filename)：以 JSON 格式导出数据到文本文件中。

3. 创建测试对象

pd.DataFrame(np.random.rand(20,5))：创建 20 行 5 列的、由随机数组成的 DataFrame 对象。

pd.Series(my_list)：从可迭代对象 my_list 中创建一个 Series 对象。

df.index = pd.date_range('1900/1/30', periods = df.shape[0])：增加一个日期索引。

4. 查看、检查数据

df.head(n)：查看 DataFrame 对象的前 *n* 行。

df.tail(n)：查看 DataFrame 对象的最后 *n* 行。

df.shape()：查看行数和列数。

http://df.info()：查看索引、数据类型和内存信息。

df.describe()：查看数值型列的汇总统计。

s.value_counts(dropna=False)：查看 Series 对象的唯一值和计数。

df.apply(pd.Series.value_counts)：查看 DataFrame 对象中每一列的唯一值和计数。

5. 数据选取

df[col]：根据列名，以 Series 的形式返回列。

df[[col1, col2]]：以 DataFrame 的形式返回多列。

s.iloc[0]：按位置选取数据。

s.loc['index_one']：按索引选取数据。

df.iloc[0,:]：返回第一行。

df.iloc[0,0]：返回第一列的第一个元素。

6. 数据清理

df.columns = ['a','b','c']：重命名列名。

pd.isnull()：检查 DataFrame 对象中的空值，并返回一个 Boolean 数组。

pd.notnull()：检查 DataFrame 对象中的非空值，并返回一个 Boolean 数组。

df.dropna()：删除所有包含空值的行。

df.dropna(axis=1)：删除所有包含空值的列。

df.dropna(axis=1,thresh=n)：保留至少有 *n* 个非空值的行。

df.fillna(x)：用 x 替换 DataFrame 对象中的所有空值。

s.astype(float)：将 Series 中的数据类型更改为 float 类型。

s.replace(1,'one')：用'one'代替所有等于 1 的值。

s.replace([1,3],['one','three'])：用'one'代替 1，用'three'代替 3。

df.rename(columns=lambda x: x + 1)：批量更改列名。

df.rename(columns={'old_name': 'new_name'})：选择性更改列名。

df.set_index('column_one')：更改索引列。

df.rename(index=lambda x: x + 1)：批量重命名索引。

7. 数据处理

df[df[col] > 0.5]：选择 col 列的值大于 0.5 的行。

df.sort_values(col1)：按照 col1 列排序数据，默认升序排列。

df.sort_values(col1, ascending=False)：按照 col1 列降序排列数据。

df.sort_values([col1,col2], ascending=[True,False])：先按 col1 列升序排列数据，再按 col2 列降序排列数据。

df.groupby(col)：返回一个按 col 列进行分组的 GroupBy 对象。

df.groupby([col1,col2])：返回一个按多列进行分组的 GroupBy 对象。

df.groupby(col1)[col2]：返回按 col1 列进行分组后，col2 列的均值。

df.pivot_table(index=col1, values=[col2,col3], aggfunc=max)：创建一个按 col1 列进行分组，并计算 col2 和 col3 的最大值的数据透视表。

df.groupby(col1).agg(np.mean)：返回按 col1 列分组的所有列的均值。

data.apply(np.mean)：对 DataFrame 中的每一列应用函数 np.mean。

data.apply(np.max,axis=1)：对 DataFrame 中的每一行应用函数 np.max。

8. 数据合并

df1.append(df2)：将 df2 中的行添加到 df1 的尾部。

df.concat([df1, df2],axis=1)：将 df2 中的列添加到 df1 的尾部。

df1.join(df2,on=col1,how='inner')：对 df1 列和 df2 列执行 SQL 形式的连接。

9. 数据统计

df.describe()：查看数据值列的汇总统计。

df.mean()：返回所有列的均值。

df.corr()：返回列与列之间的相关系数。

df.count()：返回每一列中的非空值的个数。

df.max()：返回每一列的最大值。

df.min()：返回每一列的最小值。

df.median()：返回每一列的中位数。

df.std()：返回每一列的标准差。

任务实施

1. 创建数据文件

创建 pandas_info.py 文件，并将其导入 Pandas 库，如图 4-31 所示。

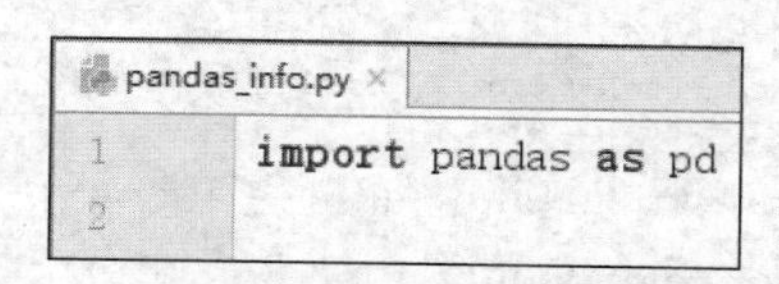

```
1  import pandas as pd
2
```

图 4-31　将文件导入 Pandas 库

2. 读取数据文件

图 4-32 所示为使用 Pandas 的 read_csv()方法读取 CSV 格式的文件，其中，delimiter 参数指定了数据文件的分隔符，encoding 参数指定了数据文件的编码，names 参数指定了数据的列索引。

```
bus_info = pd.read_csv('bus_info.csv',delimiter=',',encoding='gbk',names=['路线名称','路线类型','运行时间','票价信息',
                       '所属公司','更新时间','总里程','往线名称','往线站台详细信息','返线名称','返线站台详细信息'])
```

图 4-32　读取数据文件

3. 对数据进行去重及去空处理

如图 4-33 所示，Pandas 的 drop_duplicates()方法用于去除数据中的重复项，reset_index()方法用于还原索引为默认的整型索引（使用此方法的原因是，前面的去重或去空会清洗掉一些数据，但数据的索引仍然被保留着，导致那一行数据为空，会影响清洗操作），drop()方法用于删除数据中名为 index 的一列（axis=1 为列，默认为行），dropna()方法用于去除数据中含有任意空数据的一行数据。

```
bus_info2 = bus_info.drop_duplicates().reset_index().drop('index',axis=1)
bus_info2 = bus_info2.dropna().reset_index().drop('index',axis=1)
```

图 4-33　对数据进行去重及去空处理

4．分割、替换数据文件

由于“总里程”列的数据表述格式不清晰，因此对其做分割、替换操作。

图 4-34 所示为使用 for 循环遍历“总里程”列的所有值，并对其中的所有值进行分割、替换操作（可以通过使用 Print(bus_info['总里程'].values)方法输出所有数据）。

```
for i in range(len(bus_info2['总里程'].values)):
    if '|' in bus_info2['总里程'][i]:
        bus_info2['总里程'][i] = bus_info2['总里程'][i].split('|')[0]
        if '咨询' in bus_info2['总里程'][i]:
            bus_info2['总里程'][i] = bus_info2['总里程'][i].split('。')[0]
    elif '线路咨询' in bus_info2['总里程'][i]:
        bus_info2['总里程'][i] = bus_info2['总里程'][i].split('线路咨询')[0]
        if bus_info2['总里程'][i] == '':
            bus_info2['总里程'][i] = '没有标识'
        elif '低峰间隔' in bus_info2['总里程'][i]:
            bus_info2['总里程'][i] = bus_info2['总里程'][i].split('。')[0]
        elif '间隔' in bus_info2['总里程'][i]:
            bus_info2['总里程'][i] = bus_info2['总里程'][i].split('。')[1]
    elif '本线路|南沟村|定点班车' in bus_info2['总里程'][i] or '南沟村' in bus_info2['总里程'][i] or '定点班车' in bus_info2['总里程'][i]:
        bus_info2['总里程'][i] = '没有标识'
    elif '线路' in bus_info2['总里程'][i]:
        if '工作日' in bus_info2['总里程'][i]:
            bus_info2['总里程'][i] = bus_info2['总里程'][i].split('工作日')[0]
            bus_info2['总里程'][i] = bus_info2['总里程'][i].split('。')[1]
        elif '52' in bus_info2['总里程'][i]:
            bus_info2['总里程'][i] = bus_info2['总里程'][i].split('。')[1]
        else:
            bus_info2['总里程'][i] = bus_info2['总里程'][i].split('。')[0]
            if '全程' not in bus_info2['总里程'][i]:
                bus_info2['总里程'][i] = '没有标识'
    else:
        bus_info2['总里程'][i] = bus_info2['总里程'][i].split('。')[0]
        if '全程' not in bus_info2['总里程'][i]:
            bus_info2['总里程'][i] = '没有标识'
```

图 4-34　遍历“总里程”列的所有值并进行分割、替换操作

5．保存清洗后的数据

图 4-35 所示为使用 to_csv()方法将数据保存为 CSV 格式的文件，其中，index 属性设为 False，表示不保留索引行，header 属性设为 0，表示不保留列名（index、header 设为 False 或 0 的作用是一样的），sep 属性用于设置分隔符。保存后的文件名为 busclean_info.csv。

```
bus_info2.to_csv('busclean_info.csv',index=False, sep=',',header=0)
```

图 4-35　保存清洗后的数据

任务 4　用 OpenRefine 进行数据预处理

任务描述

（1）学习 OpenRefine 的相关基础知识。

（2）使用 OpenRefine 实现“北京公交线路信息”数据的预处理。

任务目标

（1）熟悉 OpenRefine 的相关基础知识。

（2）学会使用 OpenRefine 完成“北京公交线路信息”数据的预处理。

知识准备

OpenRefine（原名为 Google Refine）是一个用来管理杂乱数据，进行整理与扩展的工具，即使处理人员缺乏专业技术背景，其仍能够快速廉价地在一个操作界面中处理大量的数据问题。OpenRefine 可使用半自动化的方式对数据进行处理，如删除缺失值和重复值、行过滤透视、值聚类及转换、单元格拆分等。

1．运行程序

下载 openrefine-linux-3.0-rc.1.tar.gz 压缩包，将其解压到根目录中，在命令行窗口中执行“./refine”命令，启动服务，OpenRefine 会自动启动浏览器。OpenRefine Web 界面如图 4-36 所示。

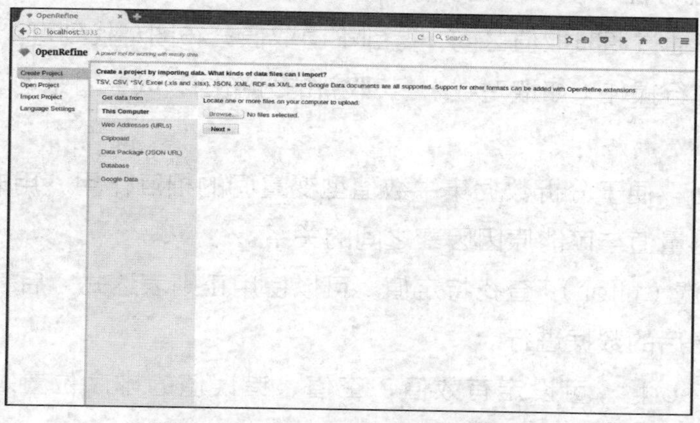

图 4-36　OpenRefine Web 界面

OpenRefine Web 界面左侧有几个选项卡，这里只介绍其中 3 个选项的功能。

（1）Create Poject（创建项目）：将载入一个数据集到 OpenRefine 中，有多种导入数据的可选形式。

① This Computer（本机）：选择本机中存储的一个文件。

② Web Addresses(URLs)（网址）：从在线资源中导入数据。

③ Clipboard（剪贴板）：通过复制粘贴的方式输入数据。

④ Google Data（Google 数据）：从 Google Sheet 或 Fusion Table 中导入数据（其类似于 Excel，但是在线的，所以需要联网）。

（2）Open Project（打开一个项目）：帮助用户定位先前创建的项目，下一次打开 OpenRefine 时，在 OpenRefine Web 界面中会出现一个已存在项目的列表，可以选择其中一个项目继续先前的工作。

（3）Import Project（导入一个项目）：可以直接导入一个已有的 OpenRefine 存档，也可以打开他人创建的 OpenRefine 项目，包含项目创建后所有的数据操作记录。

2．清洗数据

（1）找到问题数据

① 分面（facet）：分组结果显示在屏幕左栏中。OpenRefine 中有四种基本的分面：文本、数值、时间线及散布图。

文本分面：便于快速地对数据集中文本列的分布建立认知。例如，可以找到数据集中在 2019 年 11 月 1 日到 11 月 13 日之间消费额最高的城市。可按照文本出现的次数进行排序，从而判断提供的数据是否合理，对于不合理的数据可直接修改/替换。

数值/时间线分面：便于粗略了解数值型数据的分布。例如，可以检查数据集中价格的分布，可通过调节取值区间，得到不同取值分布图，找到可能有问题的数值，还可以对数值进行操作（如取对数），获取新的分布图，以更清楚地了解数值是否在合理区间内。

散布图分面：便于分析数据集中数值型变量间的相互作用，用非数学的方式来辨认某现象的测量值与可能原因因素之间的关系。

② 文本过滤（filter）：查找特定值，可以使用正则表达式，后续修改/替换等处理可仅针对过滤后的数据进行。

③ 排序（sort）：可设定有效值、空值、错误值的排序位置，易于找到问题数值。

（2）批处理

① 通用转换：去空格、更改字母大小写、更改取值类型（文本、数值、日期）等。

② 编辑/修改/替换：修改单元格值时，可选择适用于所有相同单元格。

③ 自动归类相似取值：可将数据一键合并为相同值，这是快速完成数据一致性处理的强大工具。

④ 定制文本转换：使用各种表达式（公式/函数），结果即时预览，确认正确后再执行。

⑤ 电子表格通用的功能：分列/合并、填充等。向下填充是智能的，无需指定范围，即可自动复制其值到后续空格中。

（3）所有操作历史自动保留

① 可随时通过回退（Undo/Redo）找回某个处理前的状态。要选择返回某步骤，单击该步骤的链接即可完成回退操作。

② 处理步骤可复制（例如，复制步骤用于对另一批数据做相同处理），可抽取复制操作代码，并将其粘贴到 Apply 窗口中运行（当然，也可以在运行前修改代码）。

3. 转换格式

把其他格式的数据（清单、键/值等）通过分组处理、行列转换等转换为需要的电子表格形式。

4. 增强数据

通过 API 获取外部数据，增强电子表格中的内容。

任务实施

（1）在 OpenRefine 目录中使用“./refine”命令启动 OpenRefine 服务，如图 4-37 所示。

```
hadoop@ubuntu:/usr/local/openrefine$ ./refine
You have 2747M of free memory.
Your current configuration is set to use 1400M of memory.
OpenRefine can run better when given more memory. Read our FAQ on how to allocate more memory here:
https://github.com/OpenRefine/OpenRefine/wiki/FAQ:-Allocate-More-Memory
Starting OpenRefine at 'http://127.0.0.1:3333/'
```

图 4-37　启动 OpenRefine 服务

（2）进入其 Web 操作界面，单击“浏览...”按钮，选择 bus_info.csv 文件，单击“打开”按钮，再单击“下一步”按钮，导入数据，如图 4-38 所示。

（3）进入一个新界面，在该界面中可以发现上传的 CSV 文件，如果文件出现乱码，则可以设置字符编码，应选择支持中文的编码，这里选择“GBK”编码，单击界面右上角的“新建项目”按钮，如图 4-39 所示。

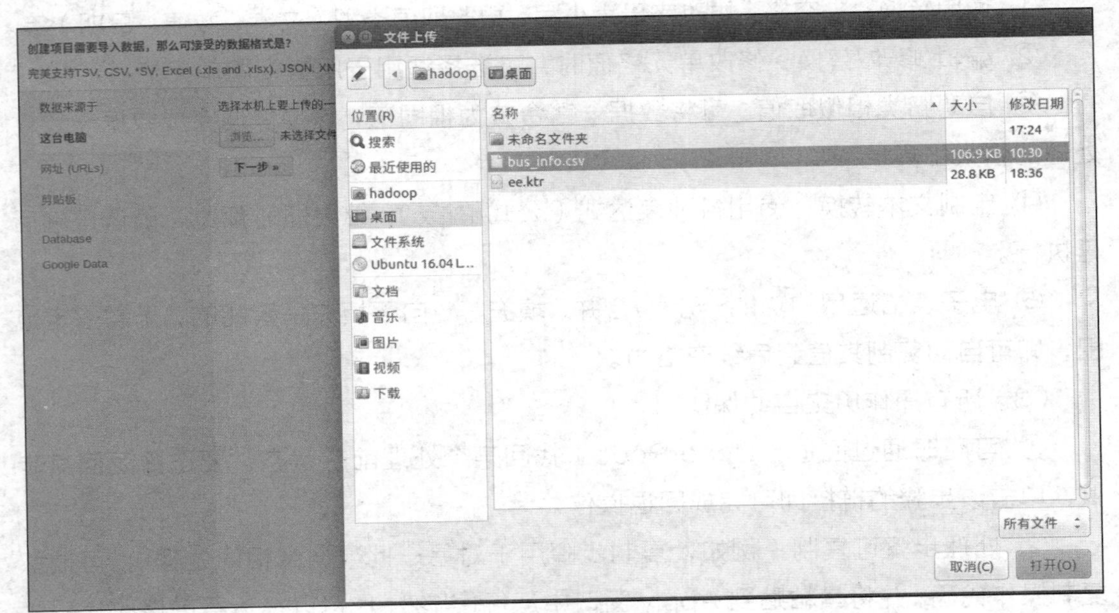

图 4-38　导入数据

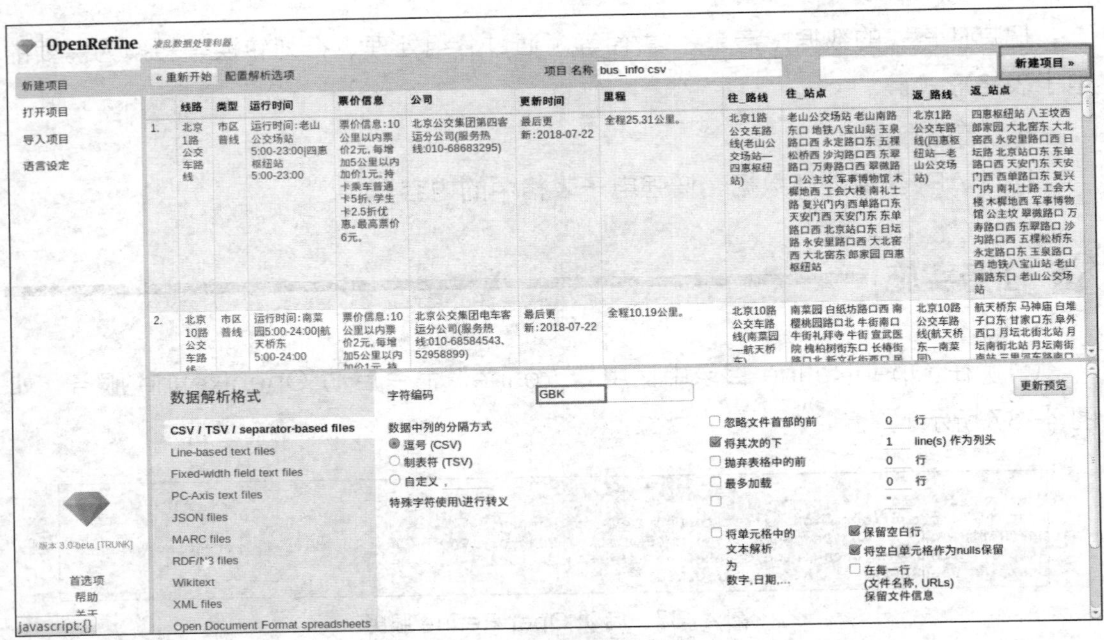

图 4-39　设置字符编码

（4）进入北京公交线路信息显示界面，在其“运行时间”列中有一些多余的信息，可将这些多余信息删除，以使数据更加简洁和直观，如图 4-40 所示。

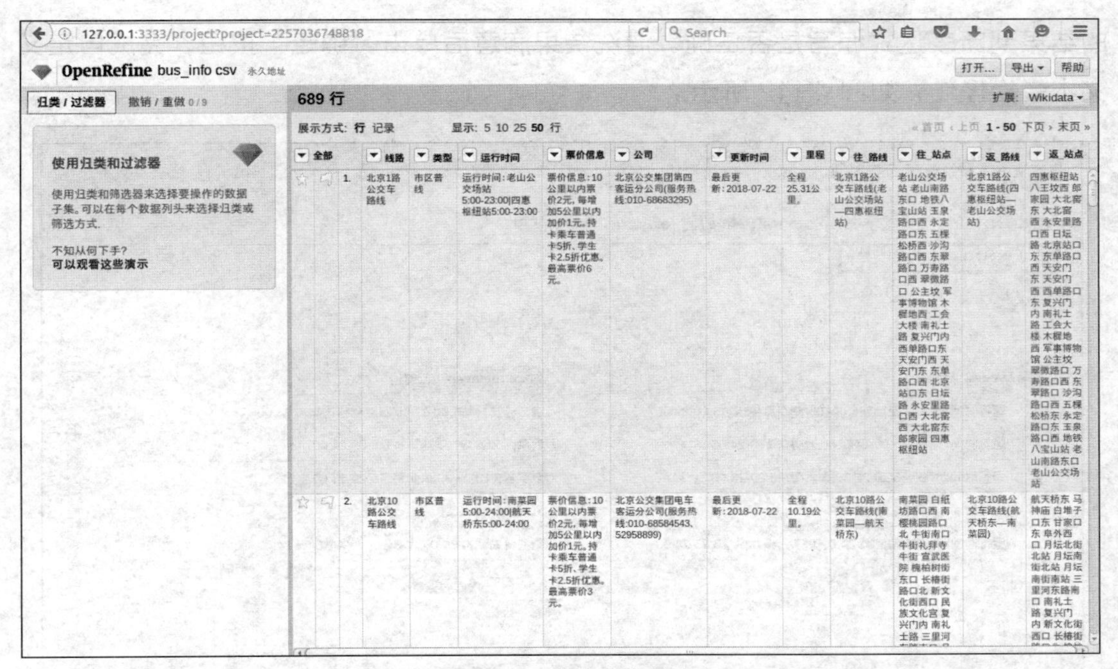

图 4-40　删除多余信息

（5）在“运行时间”下拉列表中选择“编辑单元格”中的“转换...”选项，启动转换功能，如图 4-41 所示。

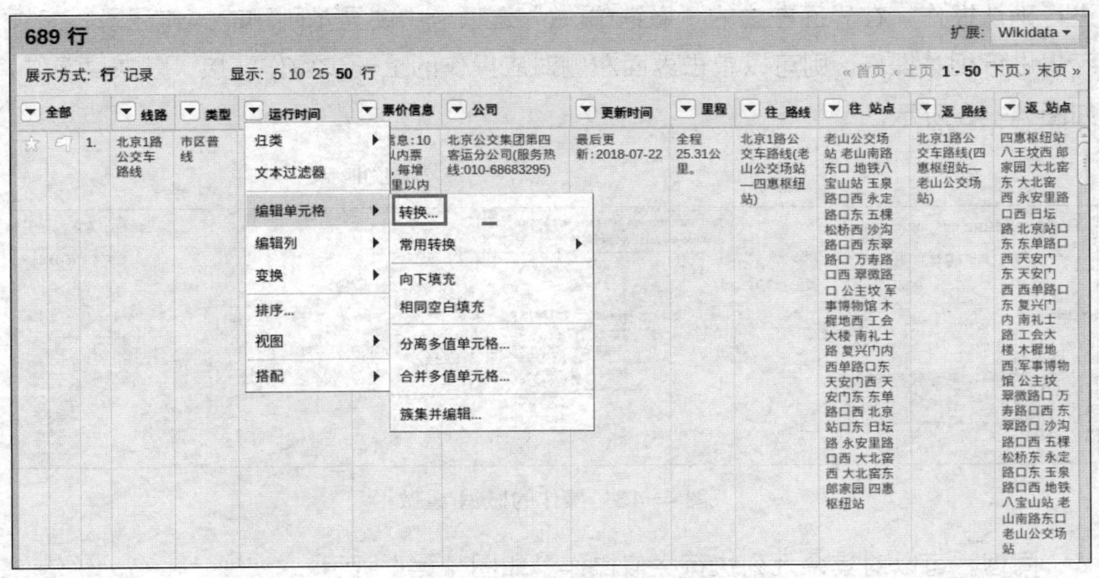

图 4-41　启动转换功能

（6）弹出“自定义文本转换于列 运行时间”对话框，在“表达式”文本框中编写表达式，去除列中“运行时间:”多余信息，编写结束后，根据“预览”选项卡中的结果判断表达式编写是否正确。清洗结果满意后单击“确定”按钮，完成自定义文本转换操作，如图 4-42 所示。

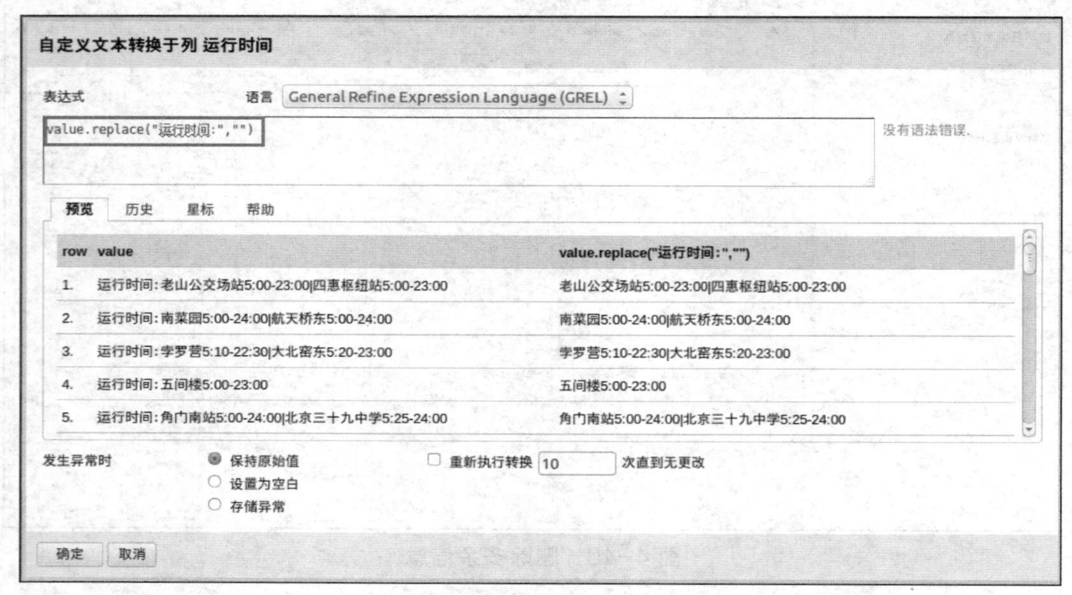

图 4-42　自定义文本转换操作

（7）界面上方弹出一个黄色通知框，通知相关操作导致改变的单元格数，再次进行确认操作。在界面左边的“撤销/重做”选项卡中会显示刚刚的操作记录，如果不想进行相关操作，则可以单击界面左侧对应操作的上一步操作链接，以恢复操作，如图 4-43 所示。

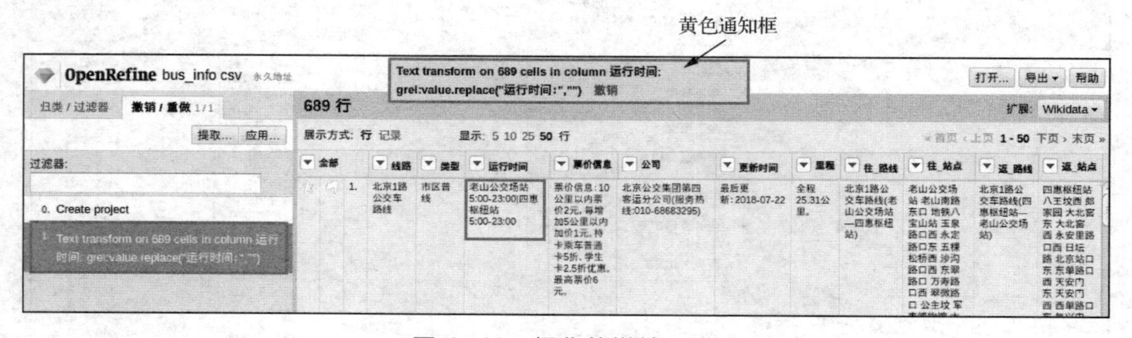

图 4-43　操作的撤销/重做

同理，可以对其余几列执行类似操作，如图 4-44 所示。

（8）操作记录及结果如图 4-45 所示。

（9）下面将“公司”列中的“服务热线”信息抽取出来并使其独立成列。在“公司”下拉列表中选择“编辑列”中的“由此列派生新列...”选项，如图 4-46 所示。

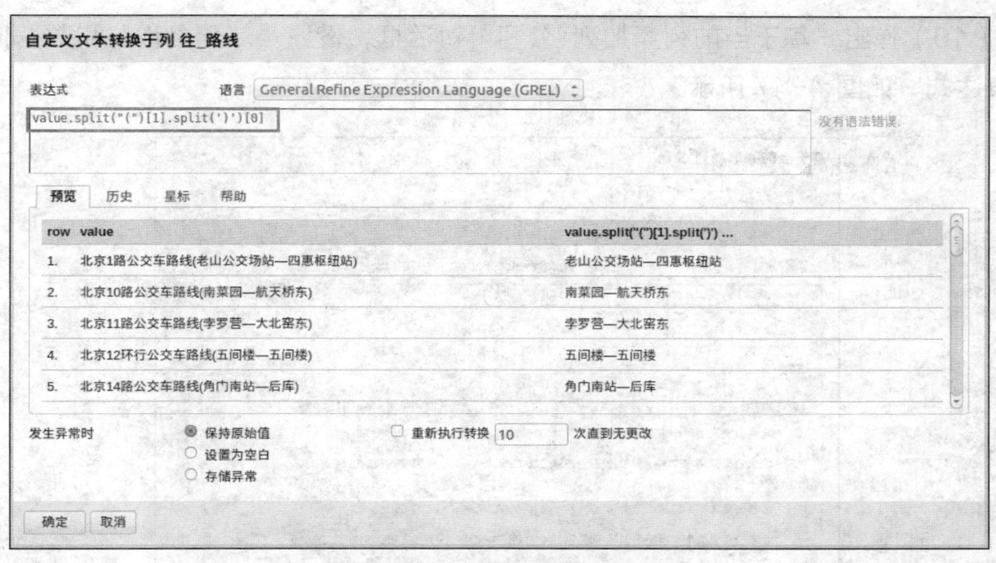

图 4-44　其他列的操作

图 4-45　操作记录及结果

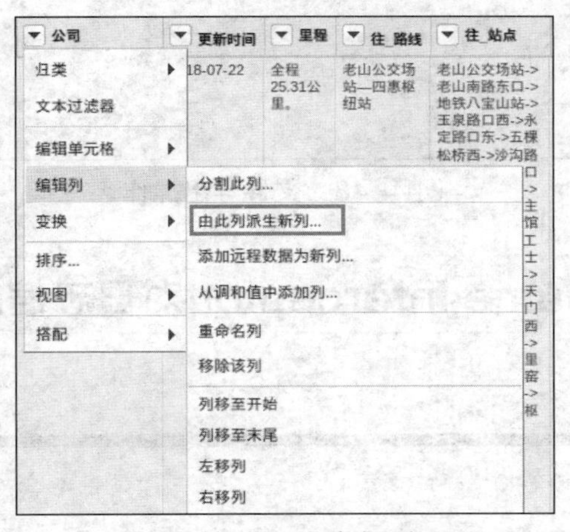

图 4-46　派生新列

（10）弹出“基于当前列添加列 公司”对话框，设置“新列名称”和数据抽取的表达式，如图 4-47 所示。

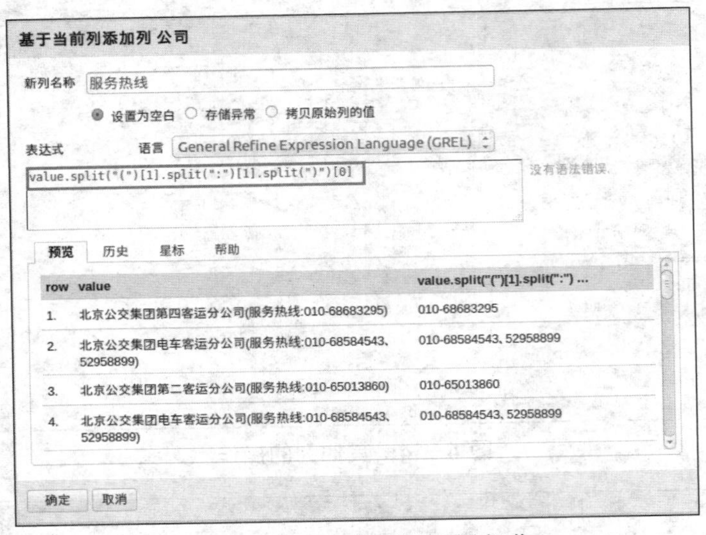

图 4-47　派生新列的相关操作

（11）操作结束后，需要将预处理后的数据导出为文件。在界面右上角单击“导出”按钮，这里选择 Excel，弹出下载对话框，单击“OK”按钮，完成文件的导出操作，如图 4-48 所示。

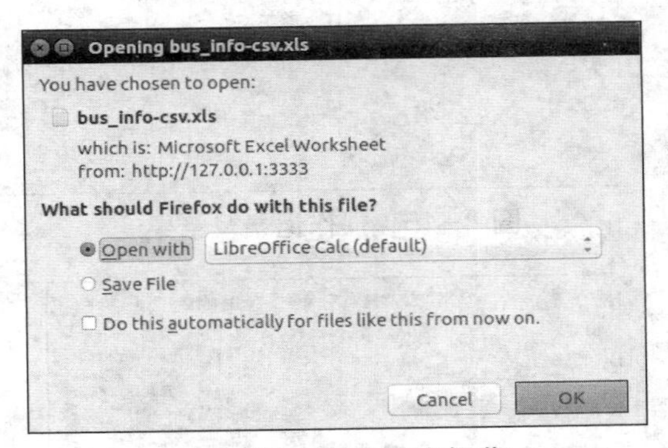

图 4-48　文件导出操作

任务 5　用 Flume Interceptor 对日志信息进行数据预处理

任务描述

（1）学习 Flume Interceptor 的相关基础知识。

（2）使用 Flume Interceptor 实现“北京公交线路信息”数据的预处理。

任务目标

（1）熟悉 Flume Interceptor 的相关基础知识。

（2）学会使用 Flume Interceptor 完成“北京公交线路信息”数据的预处理。

知识准备

有时，用户希望通过 Flume 对读取的文件进行细分存储操作。例如，将 Source 的数据按照业务类型分开存储，或者对数据进行初步筛选，丢弃或修改一些数据等。此时可以考虑使用 Flume 中的 Interceptor（拦截器），用户从 Source 读取 Events 并发送到 Sink 的时候，在 Events Header 中加入一些有用的信息，或者对 Events 的内容进行过滤，完成初步的数据清洗。事件是 Flume 传输的最小对象，用户从 Source 获取数据后，会先封装成事件，再将事件发送到 Channel，Sink 从 Channel 获取事件并进行消费。事件由头(Map<String,String> headers)和身体(body)两部分组成：Headers 部分是一个 Map，body 部分可以是 String 或 byte[]等，且 body 部分是真正存放数据的地方。

Flume 中提供了以下拦截器：时间戳拦截器（Timestamp Interceptor）、主机拦截器（Host Interceptor）、静态拦截器（Static Interceptor）、Regex 过滤拦截器（Regex Filtering Interceptor）、搜索并替换拦截器（Search and Replace Interceptor）等。

Flume 支持链式拦截，可在配置中指定构建的拦截器类的名称。在 Source 的配置中，拦截器被指定为一个以空格为间隔的列表。拦截器按照指定的顺序调用。一个拦截器返回的事件列表被传递到链中的下一个拦截器中。当一个拦截器要丢弃某些事件时，只需要在返回事件列表时不返回该事件即可。若拦截器要丢弃所有事件，则其返回一个空的事件列表即可。

通过执行“start-all.sh”命令启动 Hadoop，在任意指定目录中创建一个文件，如在/simple 目录中创建 a1.conf。

1. 时间戳拦截器

时间戳拦截器将当前时间戳（毫秒）加入 Events Header 中，key 为 timestamp，值为当前时间戳，时间戳拦截器的工作原理如图 4-49 所示。

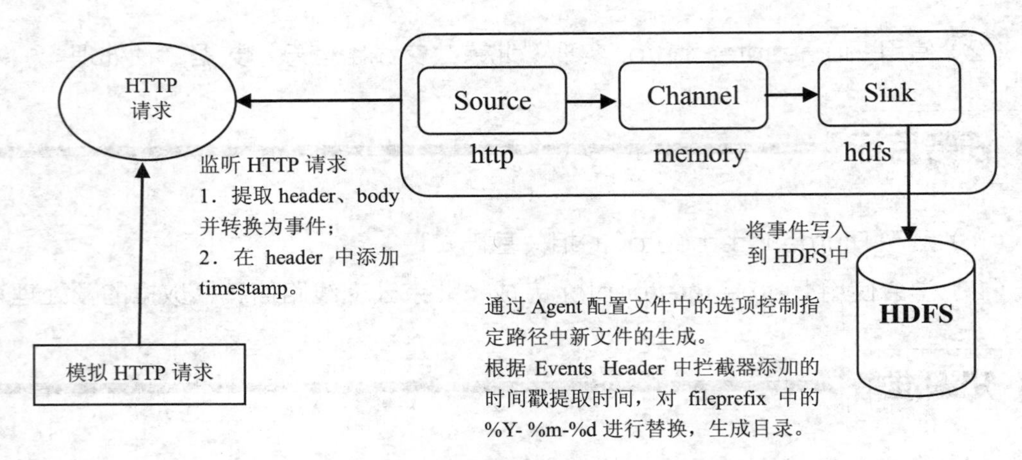

图 4-49　时间戳拦截器的工作原理

时间戳拦截器的配置参数如表 4-4 所示。

表 4-4　时间戳拦截器的配置参数

参数	默认值	描述
type		类型名称为 timestamp，也可以使用类名的全路径
preserveExisting	False	设置为 True 时，若事件中报头已经存在，则不会替换时间戳报头的值

2. 主机拦截器

主机拦截器将运行 Flume Agent 的主机名或 IP 地址加入到 Events Header 中，key 为 host（也可自定义）。主机拦截器的工作原理如图 4-50 所示。

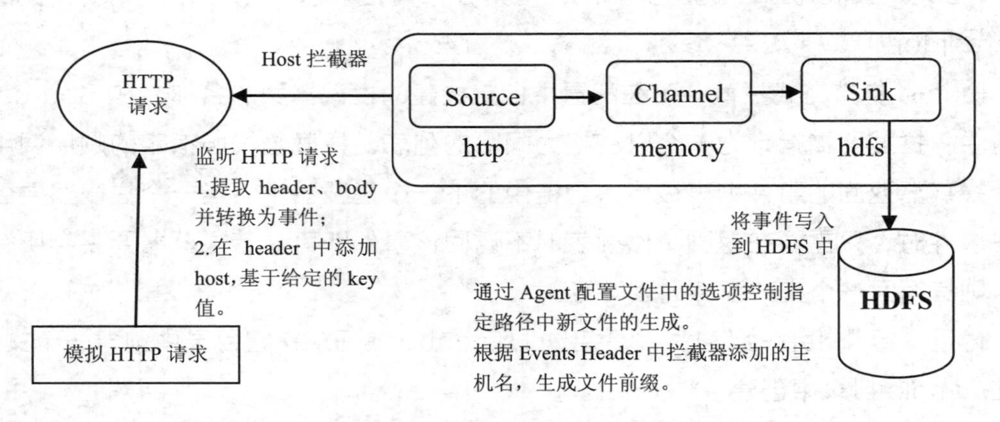

图 4-50　主机拦截器的工作原理

主机拦截器的配置参数如表 4-5 所示。

表 4-5 主机拦截器的配置参数

参数	默认值	描述
type		类型名称为 host
hostHeader	host	事件头的 key
useIP	True	如果设置为 False，则 host 键插入主机名
preserveExisting	False	设置为 True 时，若事件中报头已经存在，则不会替换 host 报头的值

3. 静态拦截器

静态拦截器用于在 Events Header 中加入一组静态的 key 和 value。静态拦截器的工作原理如图 4-51 所示。

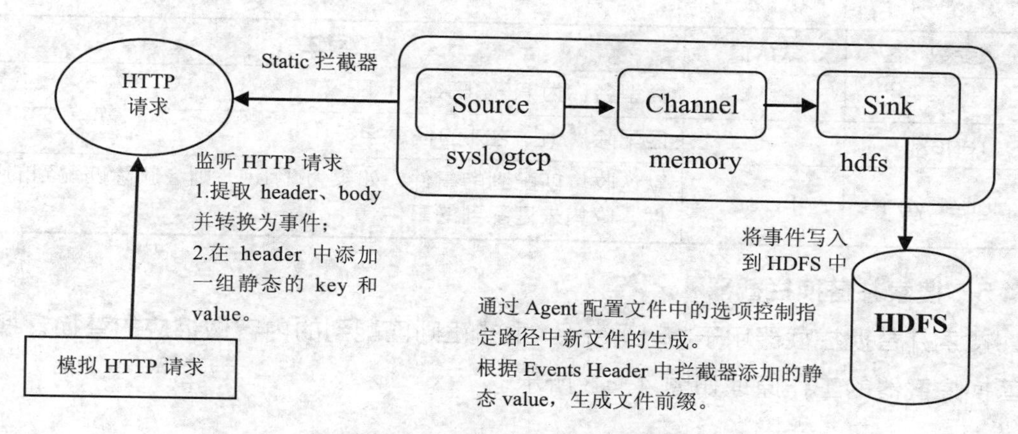

图 4-51 静态拦截器的工作原理

静态拦截器的配置参数如表 4-6 所示。

表 4-6 静态拦截器的配置参数

参数	默认值	描述
type		类型名称为 static
key	key	事件头的 key
value	value	key 对应的 value
preserveExisting	True	设置为 True 时，若事件中报头已经存在该 key，则不会替换 value 的值

4. Regex 过滤拦截器

Regex 过滤拦截器用于过滤事件，筛选出与配置的正则表达式相匹配的事件，可以用于包含事件或排除事件。Regex 过滤拦截器的工作原理如图 4-52 所示。

Regex 过滤拦截器的配置参数如表 4-7 所示。

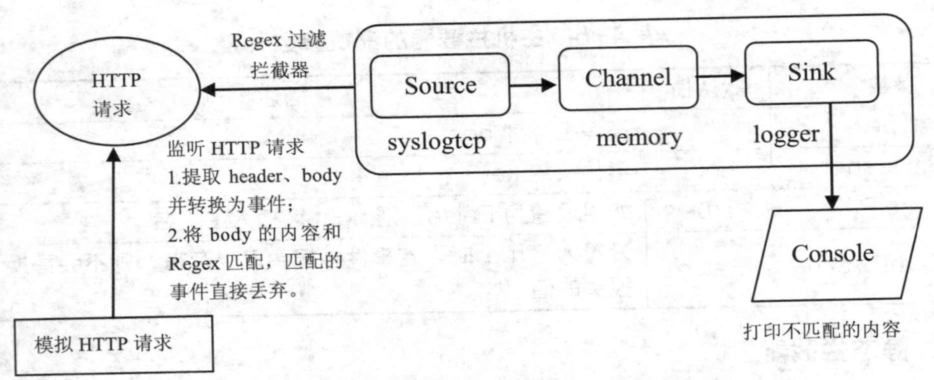

图 4-52　Regex 过滤拦截器的工作原理

表 4-7　Regex 过滤拦截器的配置参数

参数	默认值	描述
type		类型名称为 REGEX_FILTER
regex	.*	匹配除“\n”之外的任意字符
excludeEvents	False	默认收集匹配到的事件。如果为 True，则会删除匹配到的事件，收集未匹配到的事件

5. 搜索并替换拦截器

搜索并替换拦截器用于将 Events 中的正则匹配到的内容做相应的替换。搜索并替换拦截器的工作原理如图 4-53 所示。

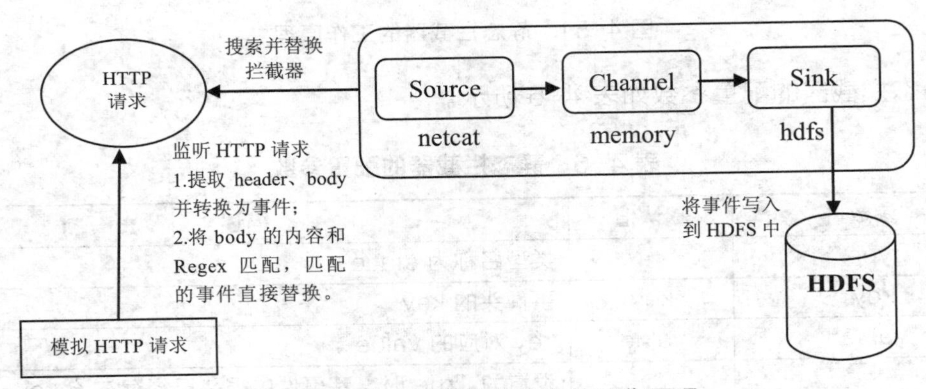

图 4-53　搜索并替换拦截器的工作原理

搜索并替换拦截器的配置参数如表 4-8 所示。

表 4-8　搜索并替换拦截器的配置参数

参数	默认值	描述
type		类型名称为 search_replace

续表

参数	默认值	描述
searchPattern		要搜索和替换的正则表达式
replaceString		要替换为的字符串
charset	utf-8	设置字符集编码

任务实施

通过执行“start-all.sh”命令启动 Hadoop，并在 Flume 安装目录的 conf 目录中创建并编写配置文件。

1．设置时间戳拦截器

将当前时间戳（毫秒）加入到 Events Header 中，key 为 timestamp，值为当前时间戳。时间戳拦截器配置文件为 mytime.conf，如图 4-54 所示。

```
# 命名 Agent 上的组件
a1.Sources = r1
a1.Sinks = k1
a1.channels = c1
# 配置 Source 信息
a1.Sources.r1.type = Syslogtcp
a1.Sources.r1.port = 50000
a1.Sources.r1.host = 192.168.27.174   #根据实际情况设置
a1.Sources.r1.channels = c1
a1.Sources.r1.interceptors = i1
a1.Sources.r1.interceptors.i1.preserveExisting= false
a1.Sources.r1.interceptors.i1.type = timestamp
# 配置 Sink 信息
a1.Sinks.k1.type = hdfs
a1.Sinks.k1.channel = c1
a1.Sinks.k1.hdfs.path =hdfs://master:9000/flume/%Y-%m-%d/%H%M
a1.Sinks.k1.hdfs.filePrefix = looklook5
a1.Sinks.k1.hdfs.fileType=DataStream
# 使用缓冲内存事件的 Channel
a1.channels.c1.type = memory
a1.channels.c1.capacity = 1000
a1.channels.c1.transactionCapacity = 100
```

图 4-54　mytime.conf 配置文件

在设置好环境变量的情况下，进入 Flume 目录，执行 Flume 命令，如图 4-55 所示。

Flume 终端启动成功，其启动信息如图 4-56 所示。

此时，打开另一个终端，通过使用“curl”命令向 50000 端口发送请求信息，使 Flume 获取生成时间戳的日志信息，如图 4-57 所示。

Flume 的返回信息如图 4-58 所示。

```
drwxr-xr-x 2 bailing bailing 4096 Aug  3 08:16 ./
drwxr-xr-x 7 bailing bailing 4096 Jul 31 17:00 ../
-rw-r--r-- 1 root    root    1619 Aug  1 03:26 agent1.conf
-rw-r--r-- 1 bailing bailing  603 Jul 31 17:17 body_case15.conf
-rw-r--r-- 1 root    root    1066 Aug  3 08:16 extra.conf
-rw-r--r-- 1 bailing bailing 1661 Sep 15  2017 flume-conf.properties.template
-rw-r--r-- 1 bailing bailing 1455 Sep 15  2017 flume-env.ps1.template
-rw-r--r-- 1 bailing bailing 1568 Sep 15  2017 flume-env.sh.template
-rw-r--r-- 1 bailing bailing 3107 Sep 15  2017 log4j.properties
-rw-r--r-- 1 root    root     942 Aug  3 08:00 myhost.conf
-rw-r--r-- 1 root    root     990 Aug  1 05:23 mytest.conf
-rw-r--r-- 1 root    root    1619 Aug  3 07:32 mytime.conf
-rw-r--r-- 1 root    root     926 Jul 31 20:30 regex_filter_case19.conf
-rw-r--r-- 1 root    root     791 Aug  3 08:13 regex_filter_case20.conf
-rw-rw-r-- 1 bailing bailing    2 Aug  1 02:45 test.txt
-rw-r--r-- 1 root    root     785 Aug  1 04:30 time.conf
-rw-r--r-- 1 root    root     943 Jul 31 19:41 time_host_case17.conf
-rw-r--r-- 1 root    root     927 Aug  1 04:49 timestamp.conf
-rw-r--r-- 1 root    root    1416 Aug  1 04:46 timetext.conf
bailing@slave1:/opt/flume/conf$ cd ..
bailing@slave1:/opt/flume$ jps
2055 Jps
bailing@slave1:/opt/flume$ flume-ng agent -c conf -f conf/mytime.conf -n a1 -Dfl
ume.root.logger=INFO,console
```

图 4-55 执行 Flume 启动命令（1）

```
)] Channel c1 connected to [k1]
2018-08-10 04:47:26,921 (conf-file-poller-0) [INFO - org.apache.flume.node.Appli
cation.startAllComponents(Application.java:137)] Starting new configuration:{ so
urceRunners:{} sinkRunners:{k1=SinkRunner: { policy:org.apache.flume.sink.Defaul
tSinkProcessor@76b8e729 counterGroup:{ name:null counters:{} } }} channels:{c1=o
rg.apache.flume.channel.MemoryChannel{name: c1}} }
2018-08-10 04:47:26,922 (conf-file-poller-0) [INFO - org.apache.flume.node.Appli
cation.startAllComponents(Application.java:144)] Starting Channel c1
2018-08-10 04:47:27,051 (lifecycleSupervisor-1-0) [INFO - org.apache.flume.instr
umentation.MonitoredCounterGroup.register(MonitoredCounterGroup.java:119)] Monit
ored counter group for type: CHANNEL, name: c1: Successfully registered new MBea
n.
2018-08-10 04:47:27,053 (lifecycleSupervisor-1-0) [INFO - org.apache.flume.instr
umentation.MonitoredCounterGroup.start(MonitoredCounterGroup.java:95)] Component
 type: CHANNEL, name: c1 started
2018-08-10 04:47:27,054 (conf-file-poller-0) [INFO - org.apache.flume.node.Appli
cation.startAllComponents(Application.java:171)] Starting Sink k1
2018-08-10 04:47:27,056 (lifecycleSupervisor-1-1) [INFO - org.apache.flume.instr
umentation.MonitoredCounterGroup.register(MonitoredCounterGroup.java:119)] Monit
ored counter group for type: SINK, name: k1: Successfully registered new MBean.
2018-08-10 04:47:27,056 (lifecycleSupervisor-1-1) [INFO - org.apache.flume.instr
umentation.MonitoredCounterGroup.start(MonitoredCounterGroup.java:95)] Component
 type: SINK, name: k1 started
```

图 4-56 Flume 启动信息（1）

```
bailing@slave1:/opt/flume/conf$ curl -X POST -d '[{"headers":{"looklook1" : "loo
klook1 isheader","looklook2": "looklook2 isheader"},"body" : "hellolooklook5"}]'
 http://192.168.27.174:50000
```

图 4-57 发送请求信息（1）

在 HDFS 中查看生成的日志文件，如图 4-59 所示。

2. 设置主机名拦截器

将运行 Flume Agent 的主机名或 IP 地址加入到 Events Header 中，key 为 host（也可自定义）。创建并编辑配置文件 myhost.conf，如图 4-60 所示。

```
bailing@slave1: /opt/flume
ogUtils.buildEvent(SyslogUtils.java:317)] Event created from Invalid Syslog data
.
2018-08-10 04:54:41,284 (New I/O worker #1) [WARN - org.apache.flume.source.Sysl
ogUtils.buildEvent(SyslogUtils.java:317)] Event created from Invalid Syslog data
.
2018-08-10 04:54:41,284 (New I/O worker #1) [WARN - org.apache.flume.source.Sysl
ogUtils.buildEvent(SyslogUtils.java:317)] Event created from Invalid Syslog data
.
2018-08-10 04:54:41,312 (SinkRunner-PollingRunner-DefaultSinkProcessor) [INFO -
org.apache.flume.sink.hdfs.HDFSDataStream.configure(HDFSDataStream.java:57)] Ser
ializer = TEXT, UseRawLocalFileSystem = false
2018-08-10 04:54:41,912 (SinkRunner-PollingRunner-DefaultSinkProcessor) [INFO -
org.apache.flume.sink.hdfs.BucketWriter.open(BucketWriter.java:251)] Creating hd
fs://master:9000/flume/2018-08-10/0454/slave1.1533902081313.tmp
2018-08-10 04:55:14,776 (hdfs-k1-roll-timer-0) [INFO - org.apache.flume.sink.hdf
s.BucketWriter.close(BucketWriter.java:393)] Closing hdfs://master:9000/flume/20
18-08-10/0454/slave1.1533902081313.tmp
2018-08-10 04:55:14,837 (hdfs-k1-call-runner-0) [INFO - org.apache.flume.sink.hd
fs.BucketWriter$8.call(BucketWriter.java:655)] Renaming hdfs://master:9000/flume
/2018-08-10/0454/slave1.1533902081313.tmp to hdfs://master:9000/flume/2018-08-10
/0454/slave1.1533902081313
2018-08-10 04:55:14,871 (hdfs-k1-roll-timer-0) [INFO - org.apache.flume.sink.hdf
s.HDFSEventSink$1.run(HDFSEventSink.java:382)] Writer callback called.
```

图 4-58　Flume 的返回信息（1）

```
bailing@slave1:/opt/flume/conf$ hdfs dfs -ls /flume/2018-08-10/0454
Found 1 items
-rw-r--r--   3 bailing supergroup        155 2018-08-10 04:55 /flume/2018-08-10/
0454/slave1.1533902081313
```

图 4-59　查看生成的日志文件（1）

```
# 命名 Agent 上的组件
a1.Sources = r1
a1.Sinks = k1
a1.channels = c1

# 配置 Source 信息
a1.Sources.r1.type = http
a1.Sources.r1.bind = master
a1.Sources.r1.port =50000
a1.Sources.r1.handler = org.apache.flume.Source.http.JSONHandler

# 配置 interceptor 信息
# 配置两个 interceptor 串联，依次作用于事件
a1.Sources.r1.interceptors = i1 i2
a1.Sources.r1.interceptors.i1.type = timestamp
a1.Sources.r1.interceptors.i1.preserveExisting = false

# 配置在 Flume Event 的头部添加“hostname”，即实际主机名
a1.Sources.r1.interceptors.i2.type = host
a1.Sources.r1.interceptors.i2.hostHeader = hostname
a1.Sources.r1.interceptors.i2.useIP = false
# 配置 Sink 信息
a1.Sinks.k1.type = hdfs
# Sink 将根据 Events Header 中的时间戳进行替换
a1.Sinks.k1.hdfs.path = hdfs://master:9000/flume/%Y-%m-%d/
# Sink 将根据 Events Header 中的 hostname 对应的 value 进行替换
a1.Sinks.k1.hdfs.filePrefix = %{hostname}
a1.Sinks.k1.hdfs.fileType = DataStream
a1.Sinks.k1.hdfs.writeFormat = Text
a1.Sinks.k1.hdfs.rollInterval = 0
a1.Sinks.k1.hdfs.rollCount = 10
a1.Sinks.k1.hdfs.rollSize = 1024000
# 将 Channel 的类型更改为 memory
a1.channels.c1.type = memory
a1.channels.c1.capacity = 1000
a1.channels.c1.transactionCapacity = 100
# 将 Source、Sink 绑定到 Channel
a1.Sinks.k1.channel = c1
a1.Sources.r1.channels = c1
```

图 4-60　myhost.conf 配置文件

在设置好配置文件的情况下，进入 Flume 目录，执行 Flume 命令，如图 4-61 所示。

```
flume-ng agent -c conf -f conf/myhost.conf -n a1 -Dflume.root.logger=INFO,console
```

图 4-61　执行 Flume 启动命令（2）

Flume 终端启动成功，其启动信息如图 4-62 所示。

```
bailing@slave1: /opt/flume
n.
2018-08-10 05:01:48,432 (lifecycleSupervisor-1-1) [INFO - org.apache.flume.instr
umentation.MonitoredCounterGroup.start(MonitoredCounterGroup.java:95)] Component
 type: CHANNEL, name: c1 started
2018-08-10 05:01:48,432 (conf-file-poller-0) [INFO - org.apache.flume.node.Appli
cation.startAllComponents(Application.java:171)] Starting Sink k1
2018-08-10 05:01:48,433 (conf-file-poller-0) [INFO - org.apache.flume.node.Appli
cation.startAllComponents(Application.java:182)] Starting Source r1
2018-08-10 05:01:48,438 (lifecycleSupervisor-1-0) [INFO - org.apache.flume.instr
umentation.MonitoredCounterGroup.register(MonitoredCounterGroup.java:119)] Monit
ored counter group for type: SINK, name: k1: Successfully registered new MBean.
2018-08-10 05:01:48,438 (lifecycleSupervisor-1-0) [INFO - org.apache.flume.instr
umentation.MonitoredCounterGroup.start(MonitoredCounterGroup.java:95)] Component
 type: SINK, name: k1 started
2018-08-10 05:01:48,501 (lifecycleSupervisor-1-3) [INFO - org.apache.flume.sourc
e.SyslogTcpSource.start(SyslogTcpSource.java:125)] Syslog TCP Source starting...
2018-08-10 05:01:48,549 (lifecycleSupervisor-1-3) [INFO - org.apache.flume.instr
umentation.MonitoredCounterGroup.register(MonitoredCounterGroup.java:119)] Monit
ored counter group for type: SOURCE, name: r1: Successfully registered new MBean
.
2018-08-10 05:01:48,550 (lifecycleSupervisor-1-3) [INFO - org.apache.flume.instr
umentation.MonitoredCounterGroup.start(MonitoredCounterGroup.java:95)] Component
 type: SOURCE, name: r1 started
```

图 4-62　Flume 启动信息（2）

此时，打开另一个终端，通过使用“curl”命令向 50000 端口发送请求信息，如图 4-63 所示。

```
bailing@slave1:/opt/flume$ curl -X POST -d '[{"headers":{}, "body" : "hellolookl
ook5"}]' http://192.168.27.174:50000
```

图 4-63　发送请求信息（2）

Flume 的返回信息如图 4-64 所示。

```
bailing@slave1: /opt/flume
ogUtils.buildEvent(SyslogUtils.java:317)] Event created from Invalid Syslog data
.
2018-08-10 05:02:52,668 (New I/O worker #1) [WARN - org.apache.flume.source.Sysl
ogUtils.buildEvent(SyslogUtils.java:317)] Event created from Invalid Syslog data
.
2018-08-10 05:02:52,669 (New I/O worker #1) [WARN - org.apache.flume.source.Sysl
ogUtils.buildEvent(SyslogUtils.java:317)] Event created from Invalid Syslog data
.
2018-08-10 05:02:52,687 (SinkRunner-PollingRunner-DefaultSinkProcessor) [INFO -
org.apache.flume.sink.hdfs.HDFSDataStream.configure(HDFSDataStream.java:57)] Ser
ializer = TEXT, UseRawLocalFileSystem = false
2018-08-10 05:02:53,043 (SinkRunner-PollingRunner-DefaultSinkProcessor) [INFO -
org.apache.flume.sink.hdfs.BucketWriter.open(BucketWriter.java:251)] Creating hd
fs://master:9000/flume/2018-08-10/0502/slave1.1533902572688.tmp
2018-08-10 05:03:24,607 (hdfs-k1-roll-timer-0) [INFO - org.apache.flume.sink.hdf
s.BucketWriter.close(BucketWriter.java:393)] Closing hdfs://master:9000/flume/20
18-08-10/0502/slave1.1533902572688.tmp
2018-08-10 05:03:24,654 (hdfs-k1-call-runner-0) [INFO - org.apache.flume.sink.hd
fs.BucketWriter$8.call(BucketWriter.java:655)] Renaming hdfs://master:9000/flume
/2018-08-10/0502/slave1.1533902572688.tmp to hdfs://master:9000/flume/2018-08-10
/0502/slave1.1533902572688
2018-08-10 05:03:24,662 (hdfs-k1-roll-timer-0) [INFO - org.apache.flume.sink.hdf
s.HDFSEventSink$1.run(HDFSEventSink.java:382)] Writer callback called.
```

图 4-64　Flume 的返回信息（2）

在 HDFS 中查看生成的日志文件，如图 4-65 所示。

```
bailing@slave1:/opt/flume$ hdfs dfs -ls /flume/2018-08-10/0502/
Found 1 items
-rw-r--r--   3 bailing supergroup        154 2018-08-10 05:03 /flume/2018-08-10/
0502/slave1.1533902572688
```

图 4-65　查看生成的日志文件（2）

3. 设置静态拦截器

创建并编辑配置文件 static.conf，如图 4-66 所示。

```
# 配置 Source 拦截器
# 命名此 Agent 上的组件
a1.Sources = r1
a1.Sinks = k1
a1.channels = c1
# 配置 Source 信息
a1.Sources.r1.type = Syslogtcp
a1.Sources.r1.port = 50000
a1.Sources.r1.host = 192.168.27.174
a1.Sources.r1.channels = c1
a1.Sources.r1.interceptors = i1
a1.Sources.r1.interceptors.i1.type = static
a1.Sources.r1.interceptors.i1.key = static_key
a1.Sources.r1.interceptors.i1.value = static_value
# 配置 Sink 信息
a1.Sinks.k1.type = logger
# 使用缓冲内存事件的 Channel
a1.channels.c1.type = memory
a1.channels.c1.capacity = 1000
a1.channels.c1.transactionCapacity = 100
# 将 Source、Sink 绑定到 Channel
a1.Sources.r1.channels = c1
a1.Sinks.k1.channel = c1
```

图 4-66　static.conf 配置文件

在设置好配置文件的情况下，进入 Flume 目录，执行 Flume 命令，如图 4-67 所示。

```
bailing@slave1:/opt/flume$ flume-ng agent -c conf -f conf/static.conf -n a1 -Dfl
ume.root.logger=INFO,console
```

图 4-67　执行 Flume 启动命令（3）

Flume 终端启动成功，其启动信息如图 4-68 所示。

此时，打开另一个终端，通过使用“curl”命令向 50000 端口发送请求信息，如图 4-69 所示。

```
bailing@slave1: /opt/flume
 }} channels:{c1=org.apache.flume.channel.MemoryChannel{name: c1}} }
2018-08-10 05:56:55,083 (conf-file-poller-0) [INFO - org.apache.flume.node.Appli
cation.startAllComponents(Application.java:144)] Starting Channel c1
2018-08-10 05:56:55,175 (lifecycleSupervisor-1-0) [INFO - org.apache.flume.instr
umentation.MonitoredCounterGroup.register(MonitoredCounterGroup.java:119)] Monit
ored counter group for type: CHANNEL, name: c1: Successfully registered new MBea
n.
2018-08-10 05:56:55,178 (lifecycleSupervisor-1-0) [INFO - org.apache.flume.instr
umentation.MonitoredCounterGroup.start(MonitoredCounterGroup.java:95)] Component
 type: CHANNEL, name: c1 started
2018-08-10 05:56:55,179 (conf-file-poller-0) [INFO - org.apache.flume.node.Appli
cation.startAllComponents(Application.java:171)] Starting Sink k1
2018-08-10 05:56:55,185 (conf-file-poller-0) [INFO - org.apache.flume.node.Appli
cation.startAllComponents(Application.java:182)] Starting Source r1
2018-08-10 05:56:55,263 (lifecycleSupervisor-1-3) [INFO - org.apache.flume.sourc
e.SyslogTcpSource.start(SyslogTcpSource.java:125)] Syslog TCP Source starting...
2018-08-10 05:56:55,296 (lifecycleSupervisor-1-3) [INFO - org.apache.flume.instr
umentation.MonitoredCounterGroup.register(MonitoredCounterGroup.java:119)] Monit
ored counter group for type: SOURCE, name: r1: Successfully registered new MBean
.
2018-08-10 05:56:55,298 (lifecycleSupervisor-1-3) [INFO - org.apache.flume.instr
umentation.MonitoredCounterGroup.start(MonitoredCounterGroup.java:95)] Component
 type: SOURCE, name: r1 started
```

图 4-68　Flume 启动信息（3）

```
bailing@slave1:~$ curl -X POST -d '["headers":{'flume'},"body":{'test'}]' http:/
/192.168.80.131:50000
```

图 4-69　发送请求信息（3）

Flume 的返回信息如图 4-70 所示。

```
19/11/07 03:35:57 INFO sink.LoggerSink: Event: { headers:{Severity=0, Facility=0, flume.syslog.status=Invalid, static_ke
y=static_value} body: 50 4F 53 54 20 2F 20 48 54 54 50 2F 31 2E 31 0D POST / HTTP/1.1. }
19/11/07 03:35:57 INFO sink.LoggerSink: Event: { headers:{Severity=0, Facility=0, flume.syslog.status=Invalid, static_ke
y=static_value} body: 48 6F 73 74 3A 20 31 39 32 2E 31 36 38 2E 38 30 Host: 192.168.80 }
19/11/07 03:35:57 INFO sink.LoggerSink: Event: { headers:{Severity=0, Facility=0, flume.syslog.status=Invalid, static_ke
y=static_value} body: 55 73 65 72 2D 41 67 65 6E 74 3A 20 63 75 72 6C User-Agent: curl }
19/11/07 03:35:57 INFO sink.LoggerSink: Event: { headers:{Severity=0, Facility=0, flume.syslog.status=Invalid, static_ke
y=static_value} body: 41 63 63 65 70 74 3A 20 2A 2F 2A 0D          Accept: */*. }
19/11/07 03:35:57 INFO sink.LoggerSink: Event: { headers:{Severity=0, Facility=0, flume.syslog.status=Invalid, static_ke
y=static_value} body: 43 6F 6E 74 65 6E 74 2D 4C 65 6E 67 74 68 3A 20 Content-Length:  }
19/11/07 03:35:57 INFO sink.LoggerSink: Event: { headers:{Severity=0, Facility=0, flume.syslog.status=Invalid, static_ke
y=static_value} body: 43 6F 6E 74 65 6E 74 2D 54 79 70 65 3A 20 61 70 Content-Type: ap }
19/11/07 03:35:57 INFO sink.LoggerSink: Event: { headers:{Severity=0, Facility=0, flume.syslog.status=Invalid, static_ke
y=static_value} body: 0D                                              . }
```

图 4-70　Flume 的返回信息（3）

在 HDFS 中查看生成的日志文件，如图 4-71 所示。

```
bailing@slave1: ~
bailing@slave1:~$ hdfs dfs -ls /flume
Found 3 items
drwxrwxrwx   - bailing supergroup          0 2018-08-03 08:00 /flume/2018-08-03
drwxrwxrwx   - bailing supergroup          0 2018-08-10 05:02 /flume/2018-08-10
drwxr-xr-x   - bailing supergroup          0 2018-08-15 22:06 /flume/slave1_stat
ic_value.1458477445451.log.tmp
bailing@slave1:~$
```

图 4-71　查看生成的日志文件（3）

4. 设置 Regex 过滤拦截器

创建并编辑配置文件 regex_filter.conf，如图 4-72 所示。

```
# 命名此 Agent 上的组件
a1.Sources = r1
a1.Sinks = k1
a1.channels = c1
# 配置 Source 信息
a1.Sources.r1.type = Syslogtcp
a1.Sources.r1.port = 50000
a1.Sources.r1.host = 192.168.233.128
a1.Sources.r1.channels = c1
a1.Sources.r1.interceptors = i1
a1.Sources.r1.interceptors.i1.type =regex_filter
a1.Sources.r1.interceptors.i1.regex =^[0-9]*$
a1.Sources.r1.interceptors.i1.excludeEvents =true
# 配置 Sink 信息
a1.Sinks.k1.type = logger
# 使用缓冲内存事件的 Channel
a1.channels.c1.type = memory
a1.channels.c1.capacity = 1000
a1.channels.c1.transactionCapacity = 100
# 将 Source、Sink 绑定到 Channel
a1.Sources.r1.channels = c1
a1.Sinks.k1.channel = c1
```

图 4-72　regex_filter.conf 配置文件

在设置好配置文件的情况下，进入 Flume 目录，执行 Flume 命令，如图 4-73 所示。

```
hadoop@master:/opt/flume$ flume-ng agent -c conf -f conf/regex_filter.conf -n a1
-Dflume.root.logger=INFO,console
```

图 4-73　执行 Flume 启动命令（4）

Flume 终端启动成功，其启动信息如图 4-74 所示。

```
bailing@slave1: /opt/flume
aultSourceFactory.create(DefaultSourceFactory.java:41)] Creating instance of sou
rce r1, type http
2018-08-10 06:58:11,712 (conf-file-poller-0) [INFO - org.apache.flume.sink.Defau
ltSinkFactory.create(DefaultSinkFactory.java:42)] Creating instance of sink: k1,
 type: logger
2018-08-10 06:58:11,717 (conf-file-poller-0) [INFO - org.apache.flume.node.Abstr
actConfigurationProvider.getConfiguration(AbstractConfigurationProvider.java:116
)] Channel c1 connected to [r1, k1]
2018-08-10 06:58:11,742 (conf-file-poller-0) [INFO - org.apache.flume.node.Appli
cation.startAllComponents(Application.java:137)] Starting new configuration:{ so
urceRunners:{r1=EventDrivenSourceRunner: { source:org.apache.flume.source.http.H
TTPSource{name:r1,state:IDLE} }} sinkRunners:{k1=SinkRunner: { policy:org.apache
.flume.sink.DefaultSinkProcessor@14a9644b counterGroup:{ name:null counters:{} }
 }} channels:{c1=org.apache.flume.channel.MemoryChannel{name: c1}} }
2018-08-10 06:58:11,744 (conf-file-poller-0) [INFO - org.apache.flume.node.Appli
cation.startAllComponents(Application.java:144)] Starting Channel c1
2018-08-10 06:58:11,921 (lifecycleSupervisor-1-0) [INFO - org.apache.flume.instr
umentation.MonitoredCounterGroup.register(MonitoredCounterGroup.java:119)] Monit
ored counter group for type: CHANNEL, name: c1: Successfully registered new MBea
n.
2018-08-10 06:58:11,923 (lifecycleSupervisor-1-0) [INFO - org.apache.flume.instr
umentation.MonitoredCounterGroup.start(MonitoredCounterGroup.java:95)] Component
 type: CHANNEL, name: c1 started
```

图 4-74　Flume 启动信息（4）

此时，打开另一个终端，通过使用“curl”命令向50000端口发送请求信息，如图4-75所示。

```
bailing@slave1:/opt/flume/conf$ curl -d 'a' http://192.168.27.174:50000
^C
bailing@slave1:/opt/flume/conf$ curl -d '1222' http://192.168.27.174:50000
^C
bailing@slave1:/opt/flume/conf$ curl -d 'a222' http://192.168.27.174:50000
^C
bailing@slave1:/opt/flume/conf$
```

图4-75　发送请求信息（4）

Flume的返回信息如图4-76所示。

```
2018-08-10 06:34:04,994 (SinkRunner-PollingRunner-DefaultSinkProcessor) [INFO -
org.apache.flume.sink.LoggerSink.process(LoggerSink.java:95)] Event: { headers:{
Severity=0, Facility=0, flume.syslog.status=Invalid} body: 61
                          a }
2018-08-10 06:34:15,439 (New I/O worker #1) [WARN - org.apache.flume.source.Sysl
ogUtils.buildEvent(SyslogUtils.java:317)] Event created from Invalid Syslog data
.
2018-08-10 06:34:21,440 (New I/O worker #2) [WARN - org.apache.flume.source.Sysl
ogUtils.buildEvent(SyslogUtils.java:317)] Event created from Invalid Syslog data
.
2018-08-10 06:34:21,441 (SinkRunner-PollingRunner-DefaultSinkProcessor) [INFO -
org.apache.flume.sink.LoggerSink.process(LoggerSink.java:95)] Event: { headers:{
Severity=0, Facility=0, flume.syslog.status=Invalid} body: 61 32 32 32
                          a222 }
```

图4-76　Flume的返回信息（4）

从返回信息中可以看出，向50000端口发送的“a”“1222”，“a222”3条数据中，经过拦截器的过滤，输出的只有“a”“a222”，“1222”被认为是无效的数据，已经被拦截。

5. 设置搜索并替换拦截器

创建并编辑配置文件search.conf，如图4-77所示。

```
# 命名 Agent 上的组件
a1.sources=r1
a1.channels=c1
a1.sinks=k1

# 配置 Source 信息
a1.sources.r1.type=http
a1.sources.r1.bind=192.168.133.138
a1.sources.r1.port=50000
# 配置 interceptor 信息
a1.sources.r1.interceptors = search-replace
a1.sources.r1.interceptors.search-replace.type = search_replace
a1.sources.r1.interceptors.search-replace.searchPattern = ^[A-Za-z0-9_]+
a1.sources.r1.interceptors.search-replace.replaceString = test
a1.sources.r1.interceptors.search-replace.charset = UTF-8

# 配置 Channel
a1.channels.c1.type=memory
a1.channels.c1.capacity=1000
```

图4-77　search.conf配置文件

```
a1.channels.c1.transactionCapacity=1000

# 配置 Sink 信息
a1.sinks.k1.type=hdfs
a1.sinks.k1.hdfs.path=/netcatFlume
a1.sinks.k1.hdfs.filePrefix=%Y%m%d-
a1.sinks.k1.hdfs.writeFormat=text
a1.sinks.k1.hdfs.fileType=DataStream
a1.sinks.k1.hdfs.rollSize=10485760
a1.sinks.k1.hdfs.rollCount=0
a1.sinks.k1.hdfs.rollInterval=60
a1.sinks.k1.hdfs.useLocalTimeStamp=true

# 将 Source、Sink 绑定到 Channel
a1.sources.r1.channels=c1
a1.sinks.k1.channel=c1
```

图 4-77 search.conf 配置文件（续）

在设置好配置文件的情况下，进入 Flume 目录，执行 Flume 命令，如图 4-78 所示。

```
hadoop@master:/usr/local/flume/conf$ flume-ng agent -c . -f search.conf -n a1 -D
flume.root.logger=INFO,console
```

图 4-78 执行 Flume 启动命令（5）

Flume 终端启动成功，其启动信息如图 4-79 所示。

```
2019-11-18 10:33:15,366 (lifecycleSupervisor-1-3) [INFO - org.eclipse.jetty.util
.log.Log.initialized(Log.java:192)] Logging initialized @740ms to org.eclipse.je
tty.util.log.Slf4jLog
2019-11-18 10:33:15,627 (lifecycleSupervisor-1-3) [INFO - org.eclipse.jetty.serv
er.Server.doStart(Server.java:372)] jetty-9.4.6.v20170531
2019-11-18 10:33:15,722 (lifecycleSupervisor-1-3) [INFO - org.eclipse.jetty.serv
er.session.DefaultSessionIdManager.doStart(DefaultSessionIdManager.java:364)] De
faultSessionIdManager workerName=node0
2019-11-18 10:33:15,722 (lifecycleSupervisor-1-3) [INFO - org.eclipse.jetty.serv
er.session.DefaultSessionIdManager.doStart(DefaultSessionIdManager.java:369)] No
 SessionScavenger set, using defaults
2019-11-18 10:33:15,725 (lifecycleSupervisor-1-3) [INFO - org.eclipse.jetty.serv
er.session.HouseKeeper.startScavenging(HouseKeeper.java:149)] Scavenging every 6
60000ms
2019-11-18 10:33:15,736 (lifecycleSupervisor-1-3) [WARN - org.eclipse.jetty.secu
rity.ConstraintSecurityHandler.checkPathsWithUncoveredHttpMethods(ConstraintSecu
rityHandler.java:806)] ServletContext@o.e.j.s.ServletContextHandler@29457016{/,n
ull,STARTING} has uncovered http methods for path: /*
2019-11-18 10:33:15,750 (lifecycleSupervisor-1-3) [INFO - org.eclipse.jetty.serv
er.handler.ContextHandler.doStart(ContextHandler.java:788)] Started o.e.j.s.Serv
letContextHandler@29457016{/,null,AVAILABLE}
2019-11-18 10:33:15,794 (lifecycleSupervisor-1-3) [INFO - org.eclipse.jetty.serv
er.AbstractConnector.doStart(AbstractConnector.java:280)] Started ServerConnecto
r@34462bd{HTTP/1.1,[http/1.1]}{192.168.133.138:55555}
2019-11-18 10:33:15,794 (lifecycleSupervisor-1-3) [INFO - org.eclipse.jetty.serv
er.Server.doStart(Server.java:444)] Started @1171ms
2019-11-18 10:33:15,795 (lifecycleSupervisor-1-3) [INFO - org.apache.flume.instr
umentation.MonitoredCounterGroup.register(MonitoredCounterGroup.java:119)] Monit
ored counter group for type: SOURCE, name: r1: Successfully registered new MBean
.
2019-11-18 10:33:15,795 (lifecycleSupervisor-1-3) [INFO - org.apache.flume.instr
umentation.MonitoredCounterGroup.start(MonitoredCounterGroup.java:95)] Component
 type: SOURCE, name: r1 started
```

图 4-79 Flume 启动信息（5）

此时，打开另一个终端，通过使用“curl”命令向 50000 端口发送请求信息，如图 4-80 所示。

```
hadoop@master:/usr/local/flume/conf$ curl -X POST -d '[{"headers":{},"body":"exa
mple"}]' http://master:50000
```

图 4-80　发送请求信息（5）

Flume 的返回信息如图 4-81 所示。

```
2019-11-18 13:30:01,810 (SinkRunner-PollingRunner-DefaultSinkProcessor) [INFO -
org.apache.flume.sink.hdfs.HDFSDataStream.configure(HDFSDataStream.java:57)] Ser
ializer = TEXT, UseRawLocalFileSystem = false
2019-11-18 13:30:02,003 (SinkRunner-PollingRunner-DefaultSinkProcessor) [INFO -
org.apache.flume.sink.hdfs.BucketWriter.open(BucketWriter.java:246)] Creating /n
etcatFlume/20191118-.1574055001811.tmp
2019-11-18 13:31:02,827 (hdfs-k1-roll-timer-0) [INFO - org.apache.flume.sink.hdf
s.HDFSEventSink$1.run(HDFSEventSink.java:393)] Writer callback called.
2019-11-18 13:31:02,828 (hdfs-k1-roll-timer-0) [INFO - org.apache.flume.sink.hdf
s.BucketWriter.doClose(BucketWriter.java:438)] Closing /netcatFlume/20191118-.15
74055001811.tmp
2019-11-18 13:31:02,854 (hdfs-k1-call-runner-4) [INFO - org.apache.flume.sink.hd
fs.BucketWriter$7.call(BucketWriter.java:681)] Renaming /netcatFlume/20191118-.1
574055001811.tmp to /netcatFlume/20191118-.1574055001811
```

图 4-81　Flume 的返回信息（5）

在 HDFS 中查看生成的日志文件，如图 4-82 所示。

```
hadoop@master:/usr/local/flume/conf$ hdfs dfs -cat /netcatFlume/20191118-.157405
5001811
test
```

图 4-82　查看生成的日志文件（4）

任务 6　创新与拓展

任务描述

使用大数据处理工具 Pig 和 Kettle 等完成如下操作。

（1）使用 Kettle 预处理 MongoDB 数据库中的数据。

（2）使用 Kettle 预处理 MySQL 数据库中的数据。

（3）使用 Pig 预处理 Linux 日志数据。

任务目标

（1）学会使用 Kettle 进行大数据预处理操作。

（2）学会使用 Pig 进行大数据预处理操作。